The Demise of Environmentalism in American Law

The Demise of Environmentalism in American Law
Michael S. Greve

The AEI Press

Publisher for the American Enterprise Institute
WASHINGTON, D.C.

1996

Publication of this book was made possible by the Paul F. Oreffice Fund for Legal and Governmental Studies, American Enterprise Institute for Public Policy Research

Available in the United States from the AEI Press, c/o Publisher Resources, Inc., 1224 Heil Quaker Blvd., P.O. Box 7001, La Vergne, TN 37086-7001. Distributed outside the United States by arrangement with Eurospan, 3 Henrietta Street, London WC2E 8LU England.

Library of Congress Cataloging-in-Publication Data
Greve, Michael S.
    The demise of environmentalism in American law / Michael S. Greve.
      p. cm.
    Includes bibliographical references and index.
    ISBN 0-8447-3980-4 (cloth : alk. paper).—ISBN 0-8447-3981-2 (pbk. : alk. paper)
    1. Environmental law—United States. 2. Law reform—United States. I. Title.
KF3775.G727   1996
344.73'.046—dc20
[347.30446]   96-2191
                                                                                                              CIP

1   3   5   7   9   10   8   6   4   2

© 1996 by the American Enterprise Institute for Public Policy Research, Washington, D.C. All rights reserved. No part of this publication may be used or reproduced in any manner whatsoever without permission in writing from the American Enterprise Institute except in the case of brief quotations embodied in news articles, critical articles, or reviews. The views expressed in the publications of the American Enterprise Institute are those of the authors and do not necessarily reflect the views of the staff, advisory panels, officers, or trustees of AEI.

Printed in the United States of America

Contents

	Acknowledgments	vii
1	The Ecological Paradigm	1
	Ecology and the Law 3	
	A Bigger New Deal? 8	
	Environments—Natural and Social 11	
	The Demise of Values 18	
2	Takings	23
	From Rights to Values 25	
	From Ad Hoc Judgments to Common-Law Rules 29	
	Property, Factions, and Environmental Values 34	
	A Line in the Sand 40	
3	Standing to Sue	42
	Ecological Standing 43	
	National Wildlife Federation and the Demise of Programmatic Litigation 46	
	Defenders of Wildlife 52	
	Common-Law Analogs? 55	
	Some Connections of Standing and Takings 61	
4	Judicial Review of Environmental Regulation	64
	From Deference to Substance? 68	
	Intent and Reason 72	
	Systemic Failure 76	
5	Functional Rules for a Dysfunctional System	85
	Regulatory Failures 87	
	Policy Coordination, Trade-offs, and Paradoxes 91	
	Capture and Regulatory Obsolescence 102	

6	**ENVIRONMENTAL IDEOLOGY AND REAL-WORLD POLITICS**	107
	The Logic of Environmentalism 110	
	The Logic of the Common Law 115	
	The Ambivalence of Interest-Group Politics 118	
	The Effects of Ideology 124	
	Politics as Second-Best 132	

INDEX	139
ABOUT THE AUTHOR	147

Acknowledgments

This monograph has been a long time in the making. Due to my gainful employment with a nonprofit law firm and to other, arguably less compelling factors, I repeatedly interrupted my work on the manuscript for months at a time. With each return to work, and with additional research and deliberation, what I had originally thought of as a speculative and controversial argument began to sound more plausible. I hope this has something to do with the argument, as distinct from my subjective impressions of it.

I presented a first, rough sketch of the general argument in a 1992 conference debate sponsored by the Federalist Society. The proceedings, including my own contribution, are reprinted in volume 21 of the *Ecology Law Quarterly* (1994).

Christopher C. DeMuth of the American Enterprise Institute and David Schoenbrod of the New York Law School were kind enough to sponsor round-table discussions on early versions of the manuscript. I profited greatly from these events, and I am indebted to the organizers, their respective institutions, and the participants. Special thanks go to Don Elliott, who commented on the manuscript at the AEI discussion.

Several scholars responded generously to my requests for comments. I am particularly grateful to Jonathan Adler, Joseph Cosby, Christopher C. DeMuth, Richard A. Epstein, Leonard Leo, Michael E. Rosman, and David Schoenbrod.

Stephanie Dea, David Garland, Dawn Traverso, Robert Alt, and Jason Cooley provided valuable research assistance at various stages of the project. Louisa Coan's editorial review and comments greatly improved the final product. Ann Petty, my editor at AEI, deserves credit for her conscientiousness and her patience in coping with my extensive revisions.

As always, my biggest debt of gratitude is to Jeremy Rabkin, my friend and teacher for fourteen years. He provided two rounds of extensive and perceptive comments on the manuscript, and whatever original insights this book contains are probably his. I alone bear responsibility for the remainder.

The Demise of Environmentalism in American Law

1
The Ecological Paradigm

Our duty is to see that important legislative purposes, heralded in the halls of Congress, are not lost or misdirected in the vast hallways of the federal bureaucracy.
  Calvert Cliffs Coordinating Committee v. Atomic Energy Commission 449 F.2d 1109, 1111 (D.C. Cir. 1971)

A lot of heralded purposes ought to get lost or misdirected. . . . Yesterday's herald is tomorrow's bore.
  ANTONIN SCALIA, "The Doctrine of Standing as an Essential Element of the Separation of Powers"

This book traces the demise of environmental values in American law, more precisely, in constitutional and administrative *case* law. I have in mind not so much that the courts have become more skeptical of particular environmental regulations or that they treat environmental interest groups and regulators more harshly—and industry plaintiffs more sympathetically—than they did some two decades ago, in the heyday of the environmental era. These tendencies do characterize the case law, but they point to a broader and more profound ideological shift.

"Environmentalism" is not simply a general sentiment for clean air, endangered species, and tropical forests. In this minimalist sense, everyone is an environmentalist. At bottom environmentalism is an ideology or worldview. This "ecological paradigm" envisions a world in which everything is connected to everything else and, from this starting point, moves toward a coherent, though to my mind perverse, view of the *legal* world. Environmentalism views common-law rights—such as private property and freedom of contract—as a menace to an imperiled planet. It therefore aims to eviscerate common-law rights and to replace them with a legal regime that would organize transactions among individual citizens for a single public purpose, environmental protection. Environmentalism thus pushes toward a centralized, unlimited political scheme. To the extent that this scheme allows for "rights," they are defined and circumscribed by public purposes. Environmentalism acknowledges and even celebrates, for example, the right of "concerned citizens" to participate in environmental policy making. But no property or any other right could limit or provide a defense against the regulatory system.

To insist that environmentalism poses so fundamental a challenge to a legal order that is, after all, the foundation on which liberal democracies rest may seem an exaggeration. Most environmental activists would deny that they intend any such assault, as would practically all politicians and federal judges. But the premises of political and regulatory projects often reach much further than their advocates may recognize or be willing to admit. The perspective just outlined is not that of a handful of environmental extremists: it well-nigh *defines* the environmentalist enterprise and has profoundly influenced environmental statutes, regulatory policies, and legal scholarship—and, until about a decade ago, judicial decisions. In fact, the environmentalist logic of the law is most clearly discernible in the decisions of the 1970s and 1980s.

Over the past decade, however, the courts have come to reject the ecological paradigm. They have renounced the doctrines that flow from the paradigm, and they have reasserted harm-based, common-law–like doctrines as an organizing principle of American law.

Admittedly, this demise of environmental values is less decisive and conclusive than their original adoption. One can date the beginning of the environmental era with some precision—the 1970 Earth Day, and the passage of the National Environmental Policy Act that same year and of the Clean Air Act shortly thereafter. A handful of judicial decisions, handed down at about the same time, rang in a wholesale revolution of American law.[1] There are no comparable events to signify the *end* of the environmental era and no judicial decisions to work a comparable change.

But this is to be expected. Ideological passions never disappear without a trace. They rarely suffer a decisive rejection; with any kind of luck, they ossify into interest groups (which is generally what has happened to environmentalism).[2] Political institutions, interest groups, and bureaucracies are difficult to dislodge even when the demands that led to their creation are no longer compelling or, for that matter, even existent.

Much the same is true of courts and the law. During the environ-

---

1. They are reviewed in Richard B. Stewart, "The Reformation of American Administrative Law," *Harvard Law Review*, vol. 88 (1975), pp. 1667–813.

2. For early testimony see Andrew S. McFarland, *Public Interest Lobbies: Decision-Making on Energy* (Washington, D.C.: American Enterprise Institute, 1976), esp. pp. 4–7. A more recent, critical appraisal is Robert Gottlieb, *Forcing the Spring: The Transformation of the American Environmental Movement* (Washington, D.C.: Island Press, 1993), pp. 117–61.

mental era, federal courts established new rights—to participate and to sue; to have clean water and clean air. It takes a few appellate decisions to establish such rights, usually under the guise of extending existing ones; it takes a more concerted and sustained effort to revoke newly minted rights. The invention of environmental rights was the work of a few entrepreneurial litigators and judges; the curtailment of these rights would require a disciplined judiciary, with a clear sense of purpose, to prevail over an entrenched cadre of environmental plaintiffs and interest groups that do not readily take no for an answer and continue to enjoy substantial political support. In short, and as a result, the ecological paradigm continues to exert considerable force.

But the courts have made as much of a coherent effort to reverse the doctrines of the environmental era as one can realistically expect. Most important, the courts have deliberately and explicitly rejected the ecological paradigm *as a matter of principle.* This development is no less remarkable or significant than the judiciary's original embrace of the environmentalist perspective: it indicates a seismic shift in the intellectual ground on which environmental politics rests.

Much larger, more visible political changes may soon follow. After twenty-five years of almost uninterrupted regulatory activism, Congress is now contemplating a serious curtailment of environmental regulation. Pending bills would substantially expand the government's duty to compensate private property owners for economic losses incurred as a result of environmental regulation; other bills would subject environmental regulations to rigorous cost-benefit and risk-assessment requirements. This legislative about-face, too, explicitly rejects environmentalist premises. It was prompted in part by the judiciary's demise of the ecological paradigm, and it reflects the same intellectual disposition.

In terms of policy outcomes and real-world effects, the legislative reforms now under consideration in Congress are bound to reverberate much further than anything the courts have done or could have done. But the courts rejected the ecological paradigm when Congress was still passing ambitious environmental laws such as the 1990 Clean Air Act Amendments, a convoluted, 600-page enactment that carries environmentalist presumptions to extremes. Thus, an examination of the case law sheds light not only on the logic and the ambitious nature of the ecological paradigm but also on the reasons for and the origins of its demise.

Ecology and the Law

Environmentalism views the world as an infinitely complex, interdependent, and fragile place. On spaceship earth, small events may have large, unforeseen consequences; when a butterfly in China flaps its wings, the effects may ripple through the entire ecosystem. "Everything is connected to everything else"—formulated by Barry Commoner in 1971 as the first law of ecology—has been the central premise and operating principle of modern environmentalism.[3]

Far from being a mere slogan, the principle of universal interconnectedness has had an enormous impact on environmental law and policy. It underlies the Endangered Species Act of 1973, which presumes (or at least pretends) that every last cog in the ecological machinery is worth preserving at all costs. Similarly, if many national parks are no longer actively managed as preserves but are instead left to the forces of nature, it is because of the ecological faith that mere mortals must not meddle in closed, self-sustaining ecosystems.[4] And in areas of the world where humans cannot be excluded, their activities must be carefully monitored and controlled, since even seemingly innocent actions may produce catastrophic environmental consequences, especially in the aggregate. This perspective helps explain why environmental policies typically take the form of pervasive, command-and-control regulation.

By the same logic, the ecological view of an infinitely complex and interdependent world has worked a drastic revision of traditional, common-law notions of rights. In a world of pervasive externalities, legal relations and instruments that are modeled on private transactions seem hopelessly dysfunctional and illegitimate and must therefore be discarded.

The environmental case is clearest about private property. The common-law tradition understands property rights as a fence or boundary around a private sphere of autonomy. Central to this traditional idea of property is my right to exclude you (and all others), so long as—and because—what I do within my sphere of autonomy does not affect you. But the right to exclude loses its meaning if everything

---

3. Barry Commoner, *The Closing Circle* (New York: Knopf, 1971), is perhaps the best-known formulation of the principle of interconnectedness. But the same idea underlies, for example, Rachel Carson's enormously influential *Silent Spring* (Boston: Houghton Mifflin, 1962). The best recent critiques of the ecological perspective are Charles T. Rubin, *The Green Crusade: Rethinking the Roots of Environmentalism* (New York: Free Press, 1994), and Gregg Easterbrook, *A Moment on the Earth: The Coming Age of Environmental Optimism* (New York: Viking, 1995).

4. A harsh and powerful critique of such practices and of ecological doctrines is Alston Chase, *In a Dark Wood: The Fight over Forests and the Rising Tyranny of Ecology* (Boston: Houghton Mifflin, 1995).

I do within my boundaries affects everyone else. Environmentalists insist that no "appropriate lines [can] be drawn between uses wholly internal to one's land and those that create external harms."[5] If this is so, the presumptive case for private property becomes exceedingly weak: if the real world makes no distinctions between mine and thine, there is no reason why the law should assert such boundaries. Thus shorn of their plausibility, my assertions of rights stand revealed as a naked attempt to preserve a special privilege even in the face of the common necessity of protecting an imperiled planet. The same goes for your rights.

This logic has no limits because the externalities are everywhere. For this reason, "[t]he ecological truism that everything is connected to everything else may be the most profound challenge ever presented to established notions of property."[6] To be sure, some environmentalists view properly structured private entitlements (such as politically created and administered pollution "rights") as a potentially useful means of protecting the environment. But such entitlements are wholly unlike traditional rights; they flow from a political scheme that is revisable at will by the legislature. Free-standing rights that protect private autonomy against the state are anathema: environmentalism views them as anachronistic and archaic and eliminates them as a constitutional category.[7]

The same logic shapes environmentalism's view of who should get access to the courts. Access to the courts traditionally required standing to sue: a plaintiff had to assert some recognizable harm to himself—a violation of a common-law right or, in more recent times, so-called injury-in-fact—to be entitled to a judicial ruling on the merits of his case. Such requirements prevent officious intermeddlers and ideological plaintiffs from interfering with private transactions in which they have no actual stake and preclude them from making a federal case out of the government's decision to regulate (or not to regulate) somebody else in a particular manner. There have always been exceptions to the rule, and standing law has always been some-

---

5. Joseph L. Sax, "The Constitutional Dimensions of Property," *Loyola Los Angeles Law Review*, vol. 26 (1992), pp. 23–32.

6. Ibid.

7. Richard J. Lazarus, "Putting the Correct 'Spin' on *Lucas*," *Stanford Law Review*, vol. 45 (1993), pp. 1411, 1421. See also Joseph L. Sax, "Property Rights and the Economy of Nature: Understanding *Lucas v. South Carolina Coastal Council*," *Stanford Law Review*, vol. 45 (1993), p. 1433; and Richard B. Stewart, "PrivProp, RegProp, and Beyond," *Harvard Journal of Law and Public Policy*, vol. 13 (1990), p. 91.

what messy. But the basic principle is clear enough.

Environmentalism takes a dim view of standing rules so conceived. In a world of omnipresent externalities, what you do with your property is not your business alone but mine also (and perhaps more so, since I, among possibly many others, am at the receiving end of a use from which you alone derive a benefit). This is so regardless of your possession of what the common law viewed as a legal title; regardless of my lack thereof; and regardless of the magnitude of the harm (which may be minuscule) and of my distance from it. Just as there are no innocent parties, there are no unaffected bystanders. From this perspective, the standing question (What is it to you?) is beside the point: there is *always* an injury-in-fact, whether through tenuous chains of causation or by virtue of my thought of losing a species or wilderness, which may be painful even if I have no physical connection to the lost ecological treasure.[8] In the end, standing belongs not to harmed individuals but to trees, birds, and other bearers of ecological values. The only relevant legal inquiry is whether the "concerned citizens" who act as "private attorneys general" on their behalf are trustworthy guardians.[9]

The important point here is this: environmentalism's broad definition of injury under the standing analysis is tied up closely with the broad definition of harm that informs its analysis of property rights. The emasculation of property rights and the elimination of standing barriers that are tied to a tangible injury are opposite sides of the same coin, and both converge on a boundless definition of the government's police power.

To be sure, environmentalists and other advocates of rethinking traditional, common-law notions put the point more positively. They argue that we should cease to conceptualize autonomy from the common-law perspective of separateness and protection against force and fraud; instead, we should recognize that no man is an island and that relationship and interdependence make autonomy possible.[10] From this vantage point, we need not lament the emasculation of property

---

8. See United States v. Students to Challenge Regulatory Agency Procedures (SCRAP), 412 U.S. 669 (1973); and Lujan v. Defenders of Wildlife, 112 S. Ct. 2130, 2139 (1992) (discussing, and rejecting, environmental plaintiffs' assertion of ecosystem nexus).

9. Christopher D. Stone, "Should Trees Have Standing?" *Southern California Law Review*, vol. 45 (1972), p. 450; Joseph Vining, *Legal Identity: The Coming of Age of Public Law* (New Haven: Yale University Press, 1978).

10. See, for example, Jennifer Nedelsky, *Private Property and the Limits of American Constitutionalism* (Chicago: University of Chicago Press, 1990), p. 273.

rights. The civic, "participatory" function of citizen standing can cheerfully be offered up as more-than-adequate compensation for the loss of exclusionary rights that have, in any event, become untenable.[11] But the central point remains: the ecological paradigm substitutes collective (albeit participatory) decision making for private arrangements.

In a similar way and for the same reasons, the ecological paradigm affects statutory law, its judicial interpretation, and the judicial review of government action. If everything is connected to everything else and if we can no longer distinguish a fist in the face from the ecologist's butterfly effect, the courts are no longer capable of sorting common-law rights and wrongs from the infinite (but heretofore legally irrelevant) complexity of life. The courts no longer have a reference point for deciding which harms are worth bearing and which are not: only the legislature can make that decision. And in a sense, even the legislature cannot do so: any legal line or regulatory boundary based on harm is a common-law relic—an unwarranted ex ante judgment that only some effects count and an artificial attempt to reduce environmental complexity. Legislation pursuant to the ecological paradigm is not an exercise in line drawing but an unbounded and unconstrained commitment to environmental values.

Once the legislature has made that commitment, the courts must show extreme deference and, indeed, solicitude. Any attempt to second-guess legislative judgments is baseless and illegitimate. Courts, for example, may not give force to an industry-plaintiff's contention that a particular regulatory burden is unjustified because the plaintiff's activity, even when unregulated, would do no harm, for any such argument is based on a hoary common-law notion called causation. The judiciary's duty "to say what the law is"[12] no longer consists in the adjudication of disputes over private rights but rather in the articulation and enforcement of legislative values. Under this conception, courts are not the protectors of rights against the government; they are the handmaidens of Congress. As Jeremy Rabkin has put it, "the notion that the will of Congress must prevail is the bedrock principle of contemporary administrative law."[13]

---

11. For example, Richard J. Lazarus, "Debunking Environmental Feudalism: Promoting the Individual through the Collective Pursuit of Environmental Quality," *Iowa Law Review*, vol. 77 (1992), pp. 1739, 1773 (arguing for promoting "environmental protection in a manner that fosters individual rights by not excluding individuals from the decision making process").

12. Marbury v. Madison, 5 U.S. (1 Cranch) 137, 177 (1803).

13. Jeremy A. Rabkin, *Judicial Compulsions: How Public Law Distorts Public Policy* (New York: Basic Books, 1989), p. 80.

A Bigger New Deal?

Practicing attorneys and many legal scholars tend to think of property rights, standing, and judicial review as discrete legal questions. But as we have seen, these issues are connected by a common logic or paradigm. Throughout the environmental era that spanned the 1970s and a good part of the 1980s, that paradigm exerted a profound influence on American law. Its central assumptions were incorporated into countless environmental and health and safety statutes. Environmental statutes typically regulate the uses of private property regardless of their external effects and treat economic activity and intervention in the natural condition of things as externalities per se and as therefore subject to the police power.[14] Virtually every environmental statute contains a broadly worded provision that permits lawsuits by "any person." The same statutes abound with legislative commitments to environmental values without regard to feasibility or cost constraints. Similarly, throughout the environmental era, leading judicial decisions in environmental cases put the law on the trajectory of the ecological paradigm: toward an evisceration of property rights, toward the abolition of harm-based standing barriers, and toward an uncompromising implementation of legislative commitments to environmental protection, combined with a probing look at agency regulations that seemed to renege on those commitments.

This ecological transformation of American law and the corresponding disconnection from common-law doctrines was not quite a revolution, inasmuch as existing legal doctrines had already undermined common-law presumptions. The common law implied great confidence in the superior efficiency and legitimacy of private orderings over government planning, along with a deep distrust of interest-group politics and partial legislation (what we now call special-interest or, in the economists' parlance, rent-seeking legislation). But that confidence and distrust went by the board in the New Deal, and along with them went common-law doctrines. In the wake of the New Deal,

---

14. For example, the Surface Mining Control and Reclamation Act, 30 U.S.C. § 1201 et seq., demands both the prevention of externalities (such as downstream water pollution) and the restoration of private mining sites to their "approximate original contour," 30 U.S.C. § 1265(b)(3), regardless of externalities. Similarly, under the Clean Water Act, 33 U.S.C.A. § 1251 et seq., discharges into navigable waters are regulated under an effluent permit system that makes no systematic reference to their effects on water quality. See generally William F. Pedersen, "Turning the Tide on Water Quality," *Ecology Law Quarterly*, vol. 15 (1988), p. 69.

the courts came to accept interest-group pluralism as the theoretical base line of constitutional and administrative law.[15] The courts continued to insist, nominally at least, that government owed private owners compensation for regulation that went "too far";[16] short of wholesale expropriation, though, government was left free to adjust the benefits and burdens of public life by arranging deals among competing interest groups, without running afoul of the takings clause of the Bill of Rights. Private rights and orderings were displaced with political arrangements—typically, regulated industries under the supervision of such agencies as the Federal Communications Commission and the Securities and Exchange Commission. Predictably, the participants in these arrangements were then granted standing to protect their entitlements, whether or not they possessed common-law rights.

In the same spirit of interest-group politics, the courts abandoned what had until then been a rather probing review of legislative enactments (expressed, for example, in the interpretive rule that statutes in derogation of the common law must be narrowly construed).[17] Instead, the courts came to accept legislative orderings of formerly private markets so long as such orderings conformed to minimal notions of rationality and due process and so long as neither "preferred freedoms" (such as freedom of speech) nor "discrete and insular minorities" were affected.[18] The courts began to accord similar deference to the agencies that administered and implemented congressional orderings, on the grounds that such specialized agencies—but not generalist judges—possessed the requisite expertise.

The twin presumptions of interest-group pluralism and administrative expertise are consonant with the ecological paradigm—up to a point. Certainly the New Deal purposefully compromised common-law rights and did so long before the environmental era.[19] But these observations suggest far greater continuity than there is. In critical respects, the ecolog-

---

15. See generally Geoffrey P. Miller, "The True Story of *Carolene Products*," *Supreme Court Review* (1987), p. 397.

16. The formula comes from a pre–New Deal case, Pennsylvania Coal v. Mahon, 260 U.S. 393, 415 (1922).

17. See Securities and Exchange Comm. v. Chenery Corp., 318 U.S. 80 (1943); but see Securities and Exchange Comm. v. Chenery Corp., 332 U.S. 194 (1947). Occasionally the proposition is still cited. See United States v. Texas, 113 S. Ct. 1631, 1634 (1993).

18. U.S. v. Carolene Products Co., 304 U.S. 144, 152–53 n.4 (1938).

19. Cass R. Sunstein, *The Partial Constitution* (Cambridge: Harvard University Press, 1993), pp. 41–42 (arguing that the New Deal's central insight was that common law rights were no more neutral and prepolitical than centralized regulatory regimes).

ical paradigm marks a radical break with the New Deal.

Breaking with the common law, the New Deal displaced markets with political regimes and economic freedom to compete with (limited) freedom *from* competition. But the impetus was typically *production* values, as it had been under the common law. The New Deal carved up the world into discrete problems, markets, industries, and interests; imposed a political order that, to its architects' minds, seemed more stable, efficient, or equitable than the chaotic world of private markets; and sought to defend that order against exogenous shocks. Euphemistically, these arrangements were often called orderly markets; in purpose and operation, they were industry or interest-group cartels. Most New Deal regulations, from milk-marketing orders to minimum wage laws and collective-bargaining mandates to broadcast licensing, have this structure.

Cartels cannot be built on the rules that order and protect private markets, such as the right to exclude and freedom of contract. For this reason, New Deal regulation did not stop at the threshold of property and contract, as traditionally conceived. But cartels do need barriers, both to protect against outsiders and to police relations among the interests within. The purpose of the regulatory arrangement thus entailed, first, a natural stopping point: regulation went only so far as to protect its intended beneficiaries—the members of the cartel—from harmful effects. Typically, these effects had to be pocketbook harms that resembled a common-law injury (though they need not be anything the common lawyer would recognize as a *right*). The cartel nature of regulation entailed, second, standing rules that permit lawsuits by members of the regulated industries (for example, broadcasters)[20] but not by outsiders and members of the public, even if members of the general public are among the ostensible beneficiaries of regulation.[21] And it entailed, third, judicial deference to administrative expertise. Probing judicial review would threaten cartel arrangements, though not so much because it might be inexpert but rather because it would be disruptive.

Environmentalism's animating vision, in contrast, is not a closed (but pluralistic) world of interests and industries but a seamless web. The organization of New Deal regulation by individual industries or markets implied a judgment that the world is at least not so complex as to render such circumscribed regulatory regimes altogether ineffectual. Environmental regulation, however, cuts across the entire economy,

---

20. FCC v. Sanders Brothers Radio Station, 309 U.S. 470 (1940).

21. For example, Block v. Community Nutrition Institute, 467 U.S. 340 (1984) (denying standing to consumers challenging milk price regulation).

and what segmentation there is (typically by medium, such as air, water, and waste) is usually viewed as a lamentable concession to political realities, as distinct from a sensible organizational principle. Similarly, whereas New Deal laws instructed regulatory agencies to create and maintain stable, protective regimes, environmental laws tell the regulators that enough is never enough and prod them toward ever-escalating interventions in an environment that is infinitely complex and constantly in flux. The principal commitment of environmental regulation—in theory, if not always in reality—is not a political equilibrium among interests but the permanent pursuit of values that transcend interest-group concerns.[22] In short, environmental regulation has no principled stopping point. If we regulate stationary sources and automobiles before we regulate lawnmowers and spray cans, we do so only because the former sources are easier to regulate and because their effects are larger. But the acknowledgment of a regulatory capacity constraint does not mean that smaller or seemingly remote effects are different in kind from large, ascertainable externalities or that they can safely be ignored. *No* effect is too remote because each is connected to the system under regulation; none is too small because each may wreak havoc on the system.

These presumptions call for a style of statutory interpretation and of judicial review that is quite different from the approach of the New Deal era. There can no longer be boundaries in the form of property claims or harm-based standing barriers: no one can be an outsider, and no one an owner. There is still deference to statutes, but it is deference to the values they embody, not to the bargains they strike. And there is still deference to administrative expertise. But that deference must come to an end when an agency exercises forbearance or otherwise suggests unfaithfulness to environmental goals or mandates: such settings call for searching judicial review.[23]

Environments—Natural and Social

The ecological paradigm, then, is not simply a bigger New Deal. In form, substance, and ambition, it is unprecedented and unparalleled. We do not purport to solve the problem of hunger in America by per-

---

22. This is reflected in the formal structure of environmental statutes. See David Schoenbrod, "Goals Statutes or Rules Statutes: The Case of the Clean Air Act," *UCLA Law Review*, vol. 30 (1983), p. 740.

23. The hard-look doctrine of judicial review was developed in precisely this context. See Scenic Hudson Preservation Conf. v. Federal Power Comm'n, 354 F.2d 608 (2d Cir. 1965); Citizens to Preserve Overton Park, Inc. v. Volpe, 401 U.S. 402 (1971); Environmental Defense Fund, Inc. v. Ruckelshaus, 439 F.2d

mitting poor people to take food off supermarket shelves without paying for it; to say the least, we would find such a policy constitutionally problematic. But federal law effectively grants an analogous right of adverse possession to every endangered species: if a woodpecker decides to build its nest on your property and on the spot where you wanted to build your house, the bird wins and you lose.[24] We do not grant "any person" a right to sue the Internal Revenue Service to compel the enforcement of the tax code against other private citizens. Yet every environmental statute authorizes such citizen suits, along with attorneys' fees for prevailing plaintiffs. And we would ordinarily be uncomfortable with legislative commitments to pursue a single goal regardless of cost or any other constraint and even less comfortable with a judicial commitment to protect such impositions against claims of private rights, as opposed to the other way around. But environmental law and legislation entail precisely such arrangements.

In short, the statist doctrines that flow from the ecological paradigm have no analogs in other regulatory arenas. They have always seemed singularly plausible in the environmental context, and perhaps only in that context. For this reason, it was in the environmental area that the federal courts moved furthest toward embracing modes of legal reasoning and argumentation that are completely disconnected from traditional, common-law notions. For the same reason, the environment has held a special attraction for legal scholars who are generally sympathetic to ordering society in a more coherent fashion than seems possible within the strictures of the expansive individual rights, the fragmentation of governmental powers, and the hurly-burly of interest-group politics to which America has historically been accustomed.

The emergence of environmentalism was accompanied by a vast body of literature, most commonly known as public-law theory, which argued that modern constitutional and administrative law could no longer be understood in individualistic, common-law terms.[25] Public-law scholars view the common law as theoretically unfounded and incompatible with the requirements of an increasingly complex and in-

---

584, 598 (D.C. Cir. 1971) ("fundamental personal interests in life, health, and liberty . . . [have] always had a special claim to judicial protection").

24. Brian F. Mannix, "The Origin of Endangered Species and the Descent of Man," *The American Enterprise*, vol. 3 no. 6 (November/December 1992), p. 8. Former Justice Byron White has made the point in a similar context: Christy v. Lujan, 109 S. Ct. 317 (1989) (White, J., dissenting from denial of certiorari).

25. Among the classics of the public law literature are Ronald Dworkin, *Law's Empire* (Cambridge: Harvard University Press, 1986); Vining, *Legal Identity*; Abram Chayes, "The Role of the Judge in Public Law Litigation," *Harvard Law Review*, vol. 89 (1976), p. 1281; Owen Fiss, "The Supreme Court, 1978

terdependent world. Deeply suspicious of private orderings, they seek to conceptualize the law from the perspective of collective purposes or public values. Common-law rights tend to hinder the attainment of such purposes since they limit the substitution of private decision making with collective political arrangements. Hence, the premises of the common law (such as individual autonomy) must be relegated to the status of a public value among other equally or more attractive values.

This theory drew heavily on developments in case law and especially in environmental law. In fact, environmental concerns shaped the basic constructs of public law to a far greater extent than its advocates, intent on developing a general, across-the-board legal theory, would let on.[26] Foremost, public-law theorists have always looked to the environment as a compelling rationale for their statist presumptions. (The chain of plausibility runs from substantive environmental claims to the public-law perspective and not the other way around: no one would proffer the expansion of government power as an end in itself or, for that matter, as a plausible argument for viewing the world as a seamless web.) Similarly, it is difficult to think of any other arena where a values-centered (as opposed to rights-based) legal system would seem to hold much appeal. The public-law notion of legislation as an embodiment of consensual values seems fanciful; most traditional forms of economic regulation are readily recognized as naked preferences and wealth transfers.[27] Environmental regulation, in contrast, *can* lay some claim to incorporating important and genuinely public values that merit special judicial protection.[28]

But far from validating a general theory of public law, the prominence of substantive environmental values in public-law theory only highlights yet again the extraordinary ambition of the ecological para-

---

Term—The Forms of Justice," *Harvard Law Review*, vol. 93 (1979), p. 1; Fred Michelman, "Law's Republic," *Yale Law Journal*, vol. 97 (1988), p. 1493; and Cass R. Sunstein, "Beyond the Republican Revival," *Yale Law Journal*, vol. 97 (1988), p. 1539. A biting critique is Rabkin, *Judicial Compulsions*.

26. Robert Glicksman and Christopher H. Schroeder, "EPA and the Courts: Twenty Years of Law and Politics," *Law and Contemporary Problems*, vol. 54 (1991), pp. 249, 268–72; and Robert L. Rabin, "Federal Regulation in Historical Perspective," *Stanford Law Review*, vol. 28 (1986), pp. 1189, 1299.

27. See Cass R. Sunstein, "Naked Preferences and the Constitution," *Columbia Law Review*, vol. 84 (1984), pp. 1689, 1696.

28. Glicksman and Schroeder, "EPA and the Courts," p. 268 (throughout the 1960s and into the 1970s, the country believed environmental protection to be a "significant public value, one not reducible to the agenda of special interests intent on capturing administrative agencies"). By the same token, the market failure rationales that typically underpin environmental regulation (such

digm and the law that flows from it. Assuming that ecological complexity necessitates a radical revision of individualistic legal doctrines, there is no theoretical reason why the argument should be limited to the *natural* environment. The *social* world is every bit as complex, interdependent, and fragile as the natural world and certainly has not become less complex since the days when the common law reigned supreme. It is hard to see, then, why social complexity should not also render individualistic conceptions of rights dysfunctional.

For some theorists, social complexity has, in fact, served as a launching pad for an attack on individualistic legal doctrines.[29] The substantive commitment that drives this argument—and takes the place of ecological integrity—is social equality, and it applies most obviously and compellingly in the context of civil rights. One could argue that a private club, no matter how small and intimate, must not be allowed to exclude women or minorities, regardless of whether members of these groups in fact wish to join. Knowledge of the existence of such a club might demoralize women and minorities, signal a lack of public commitment to equality, and produce other subtle externalities.[30] Like the ecological paradigm, the argument from social complexity would eviscerate the right to exclude; confer standing on individuals who are injured by the mere thought of inequality; and legitimize the pervasive use of the police power.

For a brief period, the courts appeared to have put civil rights law on a public-law trajectory. School desegregation cases in particular exhibited the characteristics of public law: amorphous classes of beneficiaries, "rights" that defy definition, and an open-ended, one-dimensional pursuit of values regardless of adverse, self-defeating consequences (such as white flight). One could also argue with some justice that claims of a hostile work environment, which have emerged as a principal subject of employment discrimination law, owe much to public-law theory. As their name indicates, such claims are not di-

---

as pervasive externalities and inadequate information) appear far more compelling than the often transparent rationalizations of traditional, anticompetitive economic legislation.

29. See Nedelsky, *Private Property and the Limits of American Constitutionalism*, pp. 272–76.

30. If this example seems unexceptional, consider a more dramatic one: the proclivity of non-Black Americans to marry members of just about any race *except* Blacks has profound and arguably undesirable social consequences, if only by virtue of depriving young Blacks of a huge pool of otherwise eligible marriage candidates. On these grounds, one could defend a statute against racial discrimination in marriage.

rected at the employer's discrete actions toward the employee but at his sufferance of inhospitable employment conditions. The demands are for a kind of human habitat preservation, with the employer's failure to abate rude remarks and sideward glances serving as an analog for a property owner's failure—punishable under the Endangered Species Act—to manage his chunk of the ecosystem in accordance with the piping plover's interests.

In the end, however, neither academic theorists nor the courts have been prepared to apply the public-law model in full regalia to civil rights and other social contexts,[31] even while showing no such reluctance with regard to the environment. One reason for this disjunction may be that it is always easier to conceptualize a new issue in new theoretical terms than to graft a new theoretical framework onto preexisting regulatory regimes and their legal forms. (And the environment *was* a new issue when public-law doctrines began to make their mark.) Far more important is a second consideration: the regulation of social relations in light of the interconnectedness of all things would threaten social-libertarian commitments. These commitments are not simply partisan; they are intimately connected to elementary and widely shared intuitions about liberal democracy. These are held dear especially by the scholars and judges who push the ecological paradigm.

The case is clearest regarding speech, expression, and privacy. Feminist scholars and critical race theorists have urged a greater recognition of the deleterious consequences of, for example, hate speech and pornography,[32] and public-law advocates are generally receptive to such arguments. Yet no one would seriously propose that we go by the environmental playbook and set national ambient standards for pornography, with an adequate margin of moral safety for the most vulnerable member of society; issue printing permits; impose $25,000 fines per day per violation; and allow any private citizen to enforce

---

31. Professor Sunstein, for example, argues at length for government regulation that would make private media markets conform to transcendental values of "quality" and "diversity," regardless of what anybody actually wants to put on the air or on paper and what anybody actually wants to hear, see, or read; Cass R. Sunstein, *Democracy and the Problem of Free Speech* (New York: Free Press, 1993) pp. 39–43, 68–92. Here and in his earlier writings, however (and, one might add, fortunately), Sunstein shrinks from the practical implications of his proposal. See Stephen F. Williams, "Background Norms in the Regulatory State," *University of Chicago Law Review*, vol. 58 (1991), pp. 419, 427–29 (reviewing Sunstein, *After the Rights Revolution*).

32. For example, Catharine A. MacKinnon, *Only Words* (Cambridge: Harvard University Press, 1993); Mari J. Matsuda, "Public Response to Racist Speech: Considering the Victim's Story," *Michigan Law Review*, vol. 47 (1989), p. 2320.

these rules and standards. For conduct within the broad penumbras of privacy and free expression, we do not assert—as environmentalists asert about property—that "no appropriate lines can be drawn between uses wholly internal to one's own (sphere of autonomy) and those that create external harm."[33] We insist on drawing lines, even if there is considerable debate as to where, precisely, they should be drawn. The reason is not that there are "externalities" in one case but not in the other: trace levels of dioxin may do harm, but so may racial epithets. Nor are the values at stake any less important: there is no reason why the desire for sexual autonomy and integrity that lies behind the feminist campaign against pornography should command less respect than the plea for species preservation. Political and ideological commitments alone explain why we refuse to treat the social ecology as a seamless web.

In the area of civil rights, public-law advocates have been similarly reluctant to give free rein to values. The original transcendental value of civil rights law, colorblindness, was jettisoned when it appeared to hamper more egalitarian aspirations.[34] Nor has diversity—the value that has replaced colorblindness—been pursued with the same one-dimensional zeal characteristic of environmental politics. Even if diversity is widely viewed as a compelling interest, we remain squeamish about the means of achieving it.[35] While the proponents of the ecological paradigm are unapologetic about eviscerating property rights, the pursuit of diversity is still bounded by a residual shame about both the stigmatizing effects of racial preferences and the costs to elementary notions of fairness. Perhaps because these effects are not easily separated from real human beings, we treat civil rights entitlements not as fungible results of a free-floating public value but as personal rights, even if those rights are held by individuals *as members of a group*.

Certainly, the Supreme Court refused long ago to organize civil rights law in consonance with public-law conceptions. The Court tried

---

33. Recall Sax's corresponding assertion about property, quoted on page 5.

34. See, for example, Griggs v. Duke Power Co., 401 U.S. 424 (1971); United Steelworkers v. Weber, 443 U.S. 193 (1979) (permitting an affirmative action plan that allowed minority applicants to be chosen for promotion over nonminority applicants with more seniority); Fullilove v. Klutznick, 448 U.S. 448 (1980); Johnson v. Transportation Agency, Santa Clara County, 480 U.S. 616 (1987) (sustaining female preferences); Univ. of California Board of Regents v. Bakke, 438 U.S. 265 269 (1978) (plurality opinion) (permitting consideration of race as a "plus" factor in student admission to medical school); Metro Broadcasting, Inc. v. FCC, 497 U.S. 547 (1990) (sustaining racial preferences in broadcast licensing process).

35. Bakke, 438 U.S. 265 (1978).

to limit remedial powers of district courts in school desegregation cases;[36] made a point of asserting standing barriers in civil rights cases;[37] and emphasized that societal discrimination—as opposed to particularized acts or patterns—could never be a predicate for affirmative action.[38] In this manner, the Court cut off both the attenuated chains of causation and probabilistic calculations and the expansive, timeless, aspirational remedies that are the stuff of public law. Similarly, claims of a hostile work environment, for all the intriguing possibilities they offer, are quite limited in practice; legal redress is predicated on conduct that is sufficiently severe and sustained to affect the terms and conditions of the *plaintiff's* employment. In most relevant aspects, civil rights law has the pluralistic, cartel-like quality of New Deal regulation: it aims to produce stable cost- and benefit-sharing arrangements among interest groups and to protect those arrangements against outsiders.[39] On reflection, public-law advocates would not have it any other way.[40]

---

36. Milliken v. Bradley, 418 U.S. 717 (1974); Pasadena City Bd. of Educ. v. Spangler, 427 U.S. 424 (1976); Board of Educ. of Oklahoma City v. Dowell, 498 U.S. 237 (1991); Freeman v. Pitts, 112 S. Ct. 1430 (1992).

37. See, for example, Warth v. Seldin, 422 U.S. 490 (1975); Allen v. Wright, 468 U.S. 737 (1984) (parents of school children lack standing to challenge IRS policies on racially integrated schools); City of Los Angeles v. Lyons, 461 U.S. 95 (1983) (victim of police chokehold lacks standing to enjoin police from future use of this technique); Valley Forge Christian College v. Americans United for Separation of Church and State, Inc., 454 U.S. 464 (1982) (no standing for taxpayers to challenge conveyance of government property to religious organization); Simon v. Eastern Kentucky Welfare Rights Org., 426 U.S. 26 (1976) (health organizations providing services for the poor lack standing to sue secretary of the Treasury for granting favorable tax treatment to hospitals that serve the poor and indigent).

38. City of Richmond v. J.A. Croson Co., 488 U.S. 469 (1989); Adarand Constructors, Inc. v. Peña, 115 S. Ct. 2097 (1995).

39. Affirmative action programs under Title VII of the Civil Rights Act, 42 U.S.C. 2000, et seq., best illustrate such arrangements. Racial preferences are widely permissible when the costs fall on a diffuse set of outsiders, as in hiring; they are viewed as far more problematic when the costs fall on members of the cartel, as they would if preferences governed promotion or layoffs. See Wygant v. Jackson Bd. of Educ., 476 U.S. 267, 282–83 (1986) (plurality opinion). Similarly, affirmative action plans entered under consent decrees are protected against collateral attacks by unaffected parties, a term that has been interpreted broadly. See Civil Rights Act of 1991, Pub. L. No. 102-166, sec. 105–08, § 703, sec. 112, § 706(e) (1991); and generally Samuel Issacharoff, "When Substance Mandates Procedures," *Cornell Law Review*, vol. 77 (1992), pp. 189, 192.

40. Two decades ago, for example, school desegregation litigation was still

The Demise of Values

The collision of the legal doctrines that flow from the ecological paradigm with basic and widely shared intuitions about liberal democracy helps explain why those doctrines never became a paradigm for American law at large. And even in the environmental area, where federal courts did move far toward modes of legal reasoning and argumentation that are disconnected from traditional, common-law notions, the judiciary's acceptance of public law was arguably a matter of degree: although the courts followed presumptions that are consistent with an overarching theory, they never quite developed an explicit jurisprudence of public law.[41] This half-hearted and partial nature of the embrace subsequently made it easier for the courts to pull back from ecological, public-law doctrines. But the courts moved sufficiently toward the ecological paradigm that a leading public-law scholar (Cass Sunstein), writing in 1985, could state with considerable justification that the "[c]ourts have begun to develop a set of principles that amount to a public law that is independent of private law doctrines."[42]

Since then, however, this development has been reined in. Over the past decade, federal judges have expanded property owners' Fifth Amendment protection against environmental land-use regulations that effect uncompensated "takings" of private property.[43] The courts have sharply curtailed the standing of environmentalists and other

---

being held out as a paradigm of public law in action. But experience with this and other institutional reform litigation has been sufficiently sobering to prompt a rather dramatic rethinking. Professor Sunstein, for example, now propagates a so-called anticaste principle that, though ambitious in its theoretical pretensions, is extremely narrow in its practical implications. In *The Partial Constitution*, pp. 338–46, Sunstein stresses that "[a]n anticaste principle is simply beyond the capacities of the judiciary," p. 340, and that, in resolving problems of "second-class citizenship," "the traditional claims of civil rights law offer little or no help. Far better models are provided by targeted education policies, including Head Start, and by recent initiatives designed to reduce violence against women," p. 344.

41. See Michael S. Greve, "Public Law and Judicial Review," *Journal of Law and Politics*, vol. 3 (1990), pp. 559, 563.

42. Cass R. Sunstein "Interest Groups in American Public Law," *Stanford Law Review*, vol. 38 (1985), pp. 29, 74.

43. The trend has been most pronounced in the claims court and the federal circuit. See, for example, Whitney Benefits v. United States, 18 Cl. Ct. 394 (1989) (awarding $60 million for taking under Surface Mining Control Act), aff'd, 926 F.2d 1177 (Fed. Cir. 1990) cert. denied, 502 U.S. 952 (1991); Florida Rock v. United States, 8 Cl. Ct. 160 (1985) (awarding $1 million in wetlands

public interest plaintiffs to challenge agency action.[44] And, in several decisions, appellate courts have reversed and remanded environmental and safety regulations as overly stringent, unsupported by scientific evidence, or otherwise unreasonable.[45] Chapters 2, 3, and 4 discuss many of these cases and show that the courts have consciously and explicitly repudiated ecological presumptions.

It is tempting to attribute this repudiation to the increasingly conservative composition of the federal judiciary (much as one can interpret the preceding groping toward public law as a consequence of a liberal dominance on the federal bench).[46] But while judicial appointments have played a role, judicial opinions evidence a learning process that transcends partisan political considerations. The past decade has been characterized by a growing disenchantment with the results and the costs of environmental regulation. An increasing number of experts, including the Environmental Protection Agency itself, have concluded that the existing regulatory framework often addresses the

---

case), aff'd in part, vacated in part, and remanded, 791 F.2d 893 (Fed. Cir. 1986), cert. denied, 479 U.S. 1053 (1990), aff'd, 21 Cl. Ct. 161 (1990), 23 Cl. Ct. 653 (1991); vacated and remanded, 18 F.3d 1560 (Fed. Cir. 1994); Loveladies Harbor v. United States, 21 Cl. Ct. 153 (1990) (awarded full compensation to vacation home property owner), aff'd, 27 F.3d 1545 (Fed Cir. 1994); Hendler v. United States, 952 F.2d 1364 (1991). The Supreme Court's most significant takings decision in recent years are Lucas v. South Carolina Coastal Council, 112 S. Ct. 2886 (1992), discussed at length in chapter 2, and Dolan v. City of Tigard, 114 S. Ct. 2309 (1994).

44. Lujan v. National Wildlife Federation, 497 U.S. 871 (1990); Lujan v. Defenders of Wildlife, 1112 S. Ct. 2130 (1992). Both cases are discussed in chapter 3.

45. Competitive Enterprise Inst. v. National Highway Traffic Safety Admin., 956 F.2d 231 (D.C. Cir. 1992); Corrosion Proof Fittings v. Environmental Protection Agency, 947 F.2d 1201 (5th Cir. 1991); International Union, UAW v. Occupational Safety & Health Admin., 938 F.2d 1310 (D.C. Cir. 1991). See chapter 4 for a discussion of these cases.

46. Public law advocates, predictably, have resisted this interpretation, preferring instead to portray the trend toward public law as a functional adjustment to the needs of a modern society. See, for example, Stewart, "The Reformation of American Administrative Law," pp. 166–7, 1811 ("the development of an interest representation model of administrative law appears as a logical and inevitable response to . . . chang[ing] conditions"). See also Mauro Cappelletti, "Governmental and Private Advocates for the Public Interest in Civil Litigation: A Comparative Study," *Michigan Law Review*, vol. 73 (1975), pp. 793, 856, 879–80. For a critique of this view, see Michael S. Greve, "The Non-Reformation of German Administrative Law: Standing to Sue and Public Interest in West German Environmental Law," *Cornell International Law Journal*, vol. 22 (1989), pp. 197, 199.

wrong risks, at an exorbitant price.⁴⁷ At the same time, the establishment media have begun to question previously near-sacrosanct policy commitments, from Superfund to global warming prevention to automobile regulation.⁴⁸ States and municipalities have begun to protest costly environmental mandates; stringent land-use controls have led to the emergence of an increasingly influential, grass-roots property rights movement that is actively contesting the environmental movement's claim to represent the public's view on environmental matters.

The demise of environmental values in the case law was played out against this background of growing discontent. Judges, along with policy experts and regulated parties, gradually realized that there was something wrong with the regulatory system. It was not that an isolated program went spectacularly awry; rather, the regulatory system suffered from *systemic* defects, inefficiencies, and distortions. The courts concluded that the ecological paradigm, so far from remedying or ameliorating systemic regulatory failures and excesses, actually exacerbated them.

In chapter 5, I argue that this conclusion is probably correct. The ecological paradigm seeks to mimic, rather than reduce, the complexity of the world outside and attempts to trump private interests with

---

47. U.S. Environmental Protection Agency, *Reducing Risk: Setting Priorities and Strategies for Environmental Protection* (Washington, D.C.: Government Printing Office, September 1990), p. 20 (there has been "virtually no focus on relative risk and cost-effective opportunities for reducing relative risks . . . [and] little correlation between the relative risk of a particular environmental problem and the EPA budget resources dedicated to reducing it"). Some of the most perceptive critics of the existing regulatory regime are scholars of a liberal and pro-environmental disposition. See especially Cass R. Sunstein, *After the Rights Revolution* (Cambridge: Harvard University Press, 1990), and Stephen Breyer, *Breaking the Vicious Circle: Toward Effective Risk Regulation* (Cambridge: Harvard University Press, 1993).

48. See, for example, Rick Atkinson, "Europeans Investigate Why Region Is Plagued by Floods," *Washington Post,* February 4, 1995; David Shaw, "Living Scared: Why Do the Media Make Life Seem So Risky?" *Los Angeles Times,* September 11, 1994 (explaining how the media uses "cry wolf" environmental scare tactics); Stephen Budiansky, "The Doomsday Myths: By Exaggerating Environmental Dangers, Activists Have Undermined Their Credibility—and Triggered an Anti-Environmental Backlash," *U.S. News and World Report,* December 12, 1993 (environmentalists overstate evidence in order to generate concern); John Holusha, "Market Place: What Is the Price for Cleanup?" *New York Times,* March 9, 1993 (U.S. environmental program has gone "seriously awry"); Associated Press, "Safety, Clean-Air Regulations Said to Increase Car Costs," *Washington Post,* March 18, 1986; and 20/20, "Are We Scaring Ourselves to Death?" *ABC News,* April 1994.

public values. But whatever the theoretical appeal of this enterprise, its collectivist forms quickly defeated its substantive aspirations. Even an unconditional commitment to a one-dimensional value still requires judgments as to what that value might entail under given circumstances. Value-driven regulation thus requires standards and prescriptions for the minutest details of production—for every boiler, valve, and production process in every industry and factory. This system quickly produced the predictable distortions and dislocations. Regulatory commitments that did not yield an inch to the complexities of economic life generated regulatory paradoxes—that is, policies that actually retarded the attainment of the stated objectives. Environmental regulation was intended to perpetuate the pursuit of consensual values; but as Professor Richard B. Stewart, one of its principal architects, put it, the system turned into one "of Soviet-style centralized planning for the production of a clean environment."[49]

To be sure, environmental regulation is substantially more participatory than the Soviet system of shoe production. In fact, the ecological premise that there can be no unconcerned citizens makes environmental regulation more participatory, at least in aspiration, than any known regulatory system. But here again the law of unintended consequences struck back with a vengeance. Instead of making the regulatory process more deliberative and rational, maximum participation maximized the opportunities for interest-group wheedling. The public-regarding nature of environmental regulation did not necessarily prevent its use for redistributive purposes; to put it polemically, private interests could and did masquerade as public values.[50]

Rigorously pursued, the argument that the ecological paradigm produces inefficiencies and interest-group "rent seeking" leads back to the old, yet profound and oft-forgotten insight of the common law: complexity is an argument *for* private orderings, not against them.[51]

---

49. Richard B. Stewart, "Economics, Environment, and the Limits of Legal Control," *Harvard Environmental Law Review*, vol. 9 (1985), pp. 1, 9–10.

50. See generally Michael S. Greve and Fred L. Smith, *Environmental Politics: Public Costs, Private Rewards* (New York: Praeger, 1992) (examining rent seeking in environmental regulation). See also Lazarus, "Debunking Environmental Feudalism," p. 1771 ("the distribution of resources for environmental protection increasingly appears to reflect the wheeling and dealing of pork barrel politics at the national level rather than an effort to distribute resources based on a neutral assessment of the environmental threats presented"). (A footnote to this passage lists numerous references.)

51. A characteristically clever and spirited presentation of this perspective is Richard A. Epstein, *Simple Rules for a Complex World* (Cambridge: Harvard University Press, 1995).

Decentralized and flexible private arrangements are far more easily tailored to a complex world than centralized, one-size-fits-all schemes: the more thoroughly such schemes attempt to mimic complexity, the harder they will crash on the law of unintended consequences. An interdependent global ecology may be *especially* dependent on clear (if somewhat artificial) boundaries; complexity may be more manageable in private backyards than in a worldwide political commons; and trees may be better off with an owner standing next to them than with their own standing to sue.[52]

The demise of the ecological paradigm neither reflects nor portends such a dramatic reaffirmation of common-law precepts. As noted at the outset, ecological presumptions still exert a sort of gravitational pull, even if they no longer drive or dominate the law. And even the decisions that do mark a decisive break with ecological thinking fall considerably short of a full-blown resurrection of common-law doctrines. For the most part, the courts have simply ceased to view environmental issues as unique; they now subject them to the assumptions that govern all other areas of public law and policy—faith in interest-group pluralism, and an essentially managerial perspective on public policy. But even this modest return to more traditional and efficient legal doctrines is most welcome. More ambitious reforms must come, as they should, from Congress.

---

52. Compare Stone, "Should Trees Have Standing?" (answering in the affirmative) with Greve and Smith, *Environmental Politics*, pp. 177–79 (arguing that failure to allow private ownership and markets accounts for many environmental failures).

2
Takings

Over the past two decades, conflicts between traditional property rights and a dense, ever growing web of environmental land-use controls have become increasingly sharp and frequent. The constitutional battleground has been the takings clause of the Fifth Amendment, which provides that private property shall not be taken, "for public use, without just compensation." The meaning and the scope of this clause have always been a matter of controversy. But the academic debate and the fight on the ground have been particularly intense in the environmental context, where—and because—ecological presumptions push toward a wholesale evisceration of property rights.

The leading Supreme Court case on the conflict between constitutional property rights and environmental values is a 1991 decision, *Lucas v. South Carolina Coastal Council*. The petitioner, David Lucas, had purchased two vacant lots on the Isle of Palms near Charleston, South Carolina, on which he intended to build single-family homes. Subsequently, the South Carolina legislature passed the Beachfront Management Act. This enactment was ostensibly designed to serve a variety of objectives, from the promotion of tourism to the preservation of open shorelines. Chief among the stated purposes, however, was the prevention of beach erosion. Among other rules and regulations, the law enacted so-called setback lines, which had the direct effect of barring the erection of permanent structures on Lucas's land. Lucas argued that this restriction amounted to a taking of private property for which the state owed compensation.

The Supreme Court essentially agreed with Lucas's contention. In an opinion written by Justice Antonin Scalia and joined by Chief Justice William H. Rehnquist and Justices Byron R. White, Sandra Day O'Connor, and Clarence Thomas, the Court held that if a regulation deprives an owner of all economically viable use of his land, the state must pay compensation unless the regulation restricts permissible uses no further than the state's *common law* of nuisance would have permitted when the challenged regulation was enacted. The Court remanded the case to the South Carolina Supreme Court for a determination of this question of state law. The state court found that the building restrictions of the Beachfront Management Act went further than the common law of nuisance would have permitted and ruled that Lucas was therefore entitled to compensation.[1]

This result hardly justifies the enormous attention *Lucas* has received.[2] The Beachfront Management Act prohibited the construction of two single-family homes in an already developed seaside resort without any plausible explanation of how this restriction would promote the stated purpose of stemming beach erosion. If such a perfectly traditional use of property could be prohibited on such flimsy grounds without triggering compensation, it would be difficult to see what, if anything, would be left of private property. In this light, it is hardly surprising that Lucas would have won his case under a number of extant theories of the takings clause.

Nor are the broader real-world effects of the *Lucas* case particularly dramatic. By its own terms, the decision applies only to land use and real property, and even then only to the rare and special case of complete wipeouts of all economically viable uses (as opposed to partial deprivations). Because of these limitations, the decision has had only a modest impact on regulatory practices.[3]

It is rather the *reasoning* of *Lucas* that commands attention. As Joseph L. Sax has argued persuasively, Justice Scalia's majority opinion is marked by a keen recognition that the ecological paradigm—expressed in the South Carolina legislature's attempt to recruit Lucas's land into public service for the preservation of a unique ecological system—poses a fundamental challenge to traditional notions of property. The opinion squarely confronts that theoretical challenge—and decisively rejects it.[4] In so doing, moreover, the Supreme Court, for the first time in modern history, reintroduced common-law conceptions of

---

1. Lucas v. South Carolina Coastal Council, 424 S.E. 484 (1992). The case eventually settled for $850,000 in compensation and $725,000 in interest, attorneys' fees, and court costs.

2. The case has produced a flurry of legal commentary. See, for example, Richard A. Epstein, "*Lucas v. South Carolina Coastal Council*: A Tangled Web of Expectations," *Stanford Law Review*, vol. 45 (1993), p. 1369; William W. Fisher III, "The Trouble with *Lucas*," *Stanford Law Review*, vol. 45 (1993), p. 1393; and sources cited throughout this chapter.

3. See, for example, Robert H. Freilich and Elizabeth A. Garvin, "Takings after *Lucas*," in David L. Callies, ed., *After* Lucas: *Land Use Regulation and the Taking of Property without Compensation* (Chicago: American Bar Association, 1993), pp. 53–81.

4. Joseph L. Sax, "Property Rights and the Economy of Nature: Understanding *Lucas v. South Carolina Coastal Council*," *Stanford Law Review*, vol. 45 (1993), p. 1433.

property into the constitutional takings analysis. Thus, for all its practical limitations, *Lucas* signals a momentous intellectual shift.

From Rights to Values

The takings clause secures an elegant balance between public needs and private rights. On the one hand, government *may* take private property—without the owner's consent—to achieve public purposes that could otherwise be thwarted by individual citizens. (A classic example is the landowner who obstinately refuses to sell the last parcel needed for the construction of a public road.) On the other hand, the requirement of "just compensation" provides protection against two separate (albeit intimately related) dangers of democratic government: the tendency of political factions, or interest groups, to expropriate weaker factions through the legislative process and the proclivity of government to pursue popular, ostensibly public purposes by imposing the burdens not on the public but on a subset of private owners.

To this day, these basic intuitions have governed cases of physical invasions—that is, cases in which the government physically condemns or occupies private property: the courts have consistently held that such takings require compensation, even when the public purpose is manifest and the impact on the affected owners is quite trivial.[5] In this context, environmental protection is no different from any other public purpose and poses no significant constitutional problem: if the government wishes to preserve an endangered species that cannot survive without a privately owned habitat, it can do so by condemning the land and compensating the owner.

The situation is quite different with respect to so-called regulatory takings—that is, cases where the government pursues its purposes (and burdens private owners) not through physical occupation but through regulation. In such cases, the courts have over the past six decades routinely denied compensation for government regulations that would plainly have required compensation had government pursued the same purposes through physical means. (Rent controls, for example, have almost always been upheld against takings challenges.)[6] And it is in the regulatory context that environmental protection poses a unique challenge to established notions of property.

Virtually any regulation of economic activity will prohibit or restrict some heretofore lawful and profitable uses of private property

---

5. Loretto v. Teleprompter Manhattan CATV Corp., 458 U.S. 419 (1982).

6. Pennell v. City of San Jose, 485 U.S. 1 (1988); Yee v. City of Escondido, 503 U.S. 519 (1991).

and in that sense will pose a takings risk. Until the advent of environmental regulation, though, the regulatory state concerned itself principally with the inputs and outputs of production. On the input side, the New Deal regulated wages and other (heretofore contractual) terms and conditions of employment; on the output side, it attempted to control the supply of goods and, in selected areas such as meat handling or trucking, to set health and safety standards. Regulation of this type was generally unconcerned with the production process: *how* things were being produced was not thought to be terribly important, so long as potentially harmful results could be averted at the points of purchase or consumption.[7]

Although regulation of this type limits property rights and may occasionally abrogate them, it poses only a limited challenge to common-law property rights that are defined by the principle of exclusion: it draws the boundaries of property rights more narrowly, and it prohibits certain uses. Within those boundaries, however, the owner is left free to do as he pleases.

Environmentalism, in contrast, attaches enormous significance not only to the outcomes but also to the *process* of production. Everything people do within their boundaries—how they produce things and how they use their land—becomes a regulatory concern. There are no boundaries, no fences, and no innocuous uses. *Any* productive use might destroy an endangered species habitat or adversely affect a wetland; hence, should a private party own such ecologically valuable assets, all competing uses must be prohibited. Regulation of this type effectively compels private owners to preserve their property in its natural state to support the ecosystem. It is a restriction only in a technical sense; in substance, it is more akin to pressing private land into public service.

To put it the other way around: even the dyed-in-the-wool common lawyer will readily concede that I may not use my property in such a way as to harm others. Since my rights must end where those of others begin, harmful uses cannot be part of my property title to begin with. By the same logic, the government does not owe me compensation when it prohibits property uses that would harm others. But environmentalism expands this so-called nuisance exemption to the takings clause to the point where it swallows the compensation requirement. If production and land use are nuisances per se, or nearly

---

7. There are a few exceptions to this pattern; worker safety standards come to mind. But these are relatively late additions to the New Deal empire. The Occupational Health and Safety Administration, for instance, was established only in 1970.

so, regulation can never constitute a taking. From this vantage point, a payment of compensation looks downright perverse: it is a bribe to induce a private owner *not* to harm the environment—an action that he had no right to in the first place.

In the 1970s, this perspective began to shape a growing body of federal regulation and state and local land-use controls. These laws and regulations replaced more traditional regulations (such as zoning laws) as the basic stuff of takings cases. But property owners who challenged environmental regulations as takings usually lost, largely because takings jurisprudence, as it then stood, offered no firm barrier to—and, in fact, was easily adapted to—the ecological view of property rights.

Post–New Deal takings analysis partook of the Supreme Court's deferential approach to economic regulation: such regulation was sustained unless it was patently absurd or irrational. This refusal to give more than passing scrutiny to the rationality of regulation, in turn, was based on a substantive political theory: with narrow exceptions (such as freedom of speech and minority rights), the Court presumed the legislative give and take among interest groups to be generally fair and protective of private rights. In the takings area, as elsewhere in economic matters, the Court took legislative declarations of intent at face value even when the articulated public purpose was alarmingly thin and the naked intent to take from A and give to B painfully obvious.[8]

This optimistic view of interest-group politics, which is worlds apart from the profound distrust of factions and partial legislation that traditionally sustained the takings clause, conceives of property claims not as distinct rights but as little more than spruced-up interest-group claims; the focus is not on the individual aggrieved owner but on the political interest to which he belongs. Accordingly, the Supreme Court construed property claims not as clearly demarcated *rights* but as one among many competing considerations that had to be weighed in a balance. The balancing focused on questions of value, as opposed to the property owner's distinct rights and consequential losses. Time and again, the Supreme Court intoned that there was "no set formula" for the resolution of takings disputes;[9] time and again, this declaration

---

8. Takings decisions reflecting this optimistic view of interest group pluralism include Hawaiian Housing Authority v. Midkiff, 467 U.S. 229 (1984); Keystone Bituminous Coal Ass'n v. DeBenedictis, 480 U.S. 470 (1987); and Pennell v. City of San Jose, 485 U.S. 1 (1988).

9. See, for example, Penn Central Transportation Co. v. New York, 438 U.S. 104, 124 (1978) (quoting Goldblatt v. Town of Hempstead, 369 U.S. 590, 594 [1962]).

was followed by value-oriented balancing exercises that, much like mere rationality review, were routinely deployed to defeat takings claims.[10] Although the Supreme Court often revised and reformulated the takings analysis, it remained consistent in eschewing bright lines and clear rules in favor of multifactor balancing tests.[11]

Such tests are very hospitable to the ecological enterprise. Value comparisons between the purported collective benefits of regulation and an individual plaintiff's disappointed expectations have an inherent tendency to favor the government. Moreover, in the area of environmental regulation, the government's purposes seem particularly compelling, while the private owner's expectations to remain unregulated seem particularly weak. To be sure, the historic purpose of the takings clause was to prevent government from procuring public benefits by a shorter route than the constitutional way of paying for the change.[12] But while this objection seems compelling when it comes to preserving this or that landmark, it seems much less so when it comes to the restoration of entire water systems, airsheds, and ecosystems: the budgetary costs of compensating the burdened landowners would be fantastic. Somewhat perversely, then, the vast scale, the intrusiveness, and the huge regulatory and compliance costs of environmental regulation tend to weaken the position of private owners.

---

10. Richard A. Epstein, *Takings: Private Property and the Power of Eminent Domain* (Cambridge: Harvard University Press, 1985), pp. 263–66 (purported benefits of regulation are defined broadly while property owners' losses are minimized). The classic example (and to this day the bane of takings plaintiffs and their attorneys) is Penn Central Transportation Co. v. New York, 438 U.S. 104 (1978), in which the petitioner claimed that the New York Landmark Preservation Commission's denial of a building permit for a fifty-five-story office tower above Grand Central Terminal constituted a compensable taking. The Penn Central Court characterized takings jurisprudence as a decidedly "factual, ad-hoc inquiry," to be guided by such factors as the economic impact of the regulation on the property owner; the extent to which the regulation interferes with the property owner's distinct, investment-backed expectations; and the character of the governmental action. The Court minimized the impact on the owner by commingling analytically and substantively different property interests (such as the rights to the terminal and the air rights above it); found that all three balancing factors weighed in the government's favor; and rejected the petitioner's takings claim.

11. Compare, for example, Penn Central, 438 U.S. 104 (1978), with Agins v. City of Tiburon, 447 U.S. 257, 262 (1980) (regulation effects a taking either if it deprives a property owner of all economically viable use or if it fails to promote a substantial government interest; cautioning, however, that takings "necessarily requires a weighing of private and public interests").

12. Pennell, 485 U.S. at 23 (Scalia, J., dissenting).

Recent case law, however, suggests that the twin presumptions of rationality review and ad hoc balancing may no longer be operative, at least not without major qualifications. And, significantly, this trend has been driven by cases dealing with environmental land-use regulation.[13] The prominence of environmental issues in modern takings law, as well as the tenor of the decisions, indicates that the Supreme Court has come to view environmental land-use controls and the rationales that underpin them as a fundamental assault on elementary notions of property.

From Ad Hoc Judgments to Common-Law Rules

The Supreme Court's first regulatory takings decision to cast doubt on the continued validity of extreme judicial deference was *Nollan v. California Coastal Commission* (1987).[14] In *Nollan*, the Court found that the coastal commission's demand for an easement across the plaintiff's beachfront property as a condition for granting a building permit constituted a compensable taking. Justice Antonin Scalia warned that the Court was "inclined to be particularly careful" in examining a regulation "where the actual conveyance of property is made a condition to the lifting of a land use restriction."[15] Any such condition must substantially advance a legitimate state interest, and there must be a nexus—found by the Court to be missing in the case at hand—between the condition that the state seeks to impose and the purpose of the land use restriction at issue. Without such a nexus, the Court said, "the building restriction is not a valid regulation of land use but an out-and-out plan of extortion."[16]

Despite this holding and the antiregulatory tenor of Justice Scalia's opinion, *Nollan* fell well short of revolutionizing takings law. While *Nollan*'s higher standard of scrutiny often influenced the outcome of the "intense factual inquiry" called for in earlier decisions,[17] it proved highly manipulable. The decision still provided "no set formula" or clear conceptual framework for organizing the takings inquiry even in

---

13. See cases cited in chap. 1, n. 43.
14. 483 U.S. 825 (1987).
15. Ibid. at 834.
16. Ibid. at 837.
17. Nollan had a substantial impact on takings law, principally because of the heightened standard of scrutiny that the Court applied to the exaction context. See Roger and Nancie Marzulla, "Regulatory Takings in the United States Claims Court: Adjusting the Burdens That in Fairness and Equity Ought to be Borne by Society as a Whole," *Catholic University Law Review*, vol. 40 (1991), p. 3.

the limited context of exactions. Perhaps this helps explain why the Supreme Court revisited the exaction issue in 1994: in *Dolan v. Tigard*,[18] the Court reaffirmed the nexus requirement and added a further requirement of rough proportionality between the required dedication and the projected harm of a proposed land use.

Wedged between *Nollan* and its reaffirmation in *Dolan*, and of greater interest than either of these cases, is the Supreme Court's 1991 decision in *Lucas v. South Carolina Coastal Council*. The reasoning and the tenor of *Lucas* mark a sharp intellectual break not only, as noted, with the ecological paradigm but also, and more broadly, with the general assumptions of balancing and rationality as the basic building blocks of takings analysis. *Lucas* is the first regulatory takings case of the modern Supreme Court to replace ad hoc judgments and balancing tests with categorical rules and bright-line legal tests and the first to base private property rights on common-law principles and on the "historical compact recorded in the Takings Clause that has become part of our constitutional culture."[19] *Lucas* pairs this search for clear rules with a profound skepticism of legislative motives and with a resolute agnosticism concerning the purposes for which the state uses its police power. Specifically, the case rejects the idea that environmental concerns warrant special judicial recognition or a drastic revision of traditional, common-law notions of property.

Justice Scalia's majority opinion in *Lucas* is ostentatiously formalistic. After only a perfunctory acknowledgment of the ad hoc nature of takings laws, the opinion identifies

> at least two discrete categories of regulatory action as compensable without case-specific inquiry into the public interest advanced in support of the restraint. The first encompasses regulations that compel the property owner to suffer a physical "invasion" of his property. . . . The second situation in which we have found categorical treatment appropriate is where regulation denies all economically beneficial or productive uses of land.[20]

These "categorical" rules are not easily squared with the balancing exercises of the past. Thus, the statement just quoted is promptly followed by the necessary qualification that even a deprivation of all viable use will *not* constitute a taking if the regulation at issue is closely akin to the abatement or prevention of a nuisance. But even so qualified, the account of the case law is not entirely accurate. The Supreme

---

18. 114 S. Ct. 2309 (1994).
19. Lucas, 112 S. Ct. at 2900.
20. Ibid. at 2893.

Court has sometimes suggested that a state interest that is entirely unrelated to nuisance abatement may justify the uncompensated elimination of all viable use, provided the interest is of sufficient weight.[21] In other cases, the Court has suggested that the state may *not* evade compensation for total takings *even if* it engages in nuisance abatement.[22] The *Lucas* opinion, however, rejects these value-oriented balancing exercises (along with their absurd results)[23] and shoehorns them into an analysis under which the duty to compensate regulatory losses turns not on subjective comparisons of value but on the *legal nature* of the state's regulation—that is to say, its character as nuisance abatement or as something else.

The decisive question then becomes how to separate nuisance regulation from other, legitimate but compensable, exercises of the police power. Scalia rejects the possibility of drawing this line on the basis of "some objective conception of 'noxiousness'": doing so, he argues, is "difficult if not impossible" because every "benefit-conferring" regulation can also be described as "harm-preventing" (and vice versa). Legislatures will be very creative in inventing and invoking noxious use rationales, and courts will have trouble drawing the distinction between harms and benefits. And if the nuisance exemption required no more than *some* police power justification (which is necessary to sustain a regulation in any event), it would become coterminus with the police power. In less legalistic terms, if legislators could immunize laws and regulations against takings claims by shouting "global warming" on the Senate floor, they could unilaterally abrogate the Fifth Amendment.

---

21. Ibid., at 2906, 2910–12 (Blackmun, J., dissenting).

22. "[O]ur cases have never applied the nuisance exception to allow complete extinction of the value of a parcel of property." Keystone Bituminous Coal, 480 U.S. at 513 (Rehnquist, C.J., dissenting).

23. It did so despite the petitioner's contention that total takings are *always* compensable. Richard A. Epstein, "Ruminations on *Lucas v. South Carolina Coastal Council*," *Loyola Law Review*, vol. 25 (1992), pp. 1225, 1226, explains why this position is untenable:

> Suppose that the only use for a parcel of land is as a dump site for waste materials, which leach into the soil and poison underground waters. According to Lucas's theory, the State would be required to pay full value to the owner of the dump site in order to shut it down. The more intense the use of the property in this way, the greater the external threat of harm, and the more the State has to pay in order to quell it. It is only when the wipeout of the landowner's use is partial that the State may impose its restrictions on use without having to pay compensation.

The *Lucas* majority opinion portrays this logic as "self-evident" and inescapable. But it is not obvious why it should be so. Of course, legislators can cite a harm-preventing rationale for just about any regulation. But the compensation requirement would disappear in the vortex of the nuisance exemption *only* if the courts were to take legislative pronouncements at face value—if, that is, the standard of review were "rational basis" or, in Justice Scalia's colorful phrase, "a test of whether the legislature has a stupid staff."[24] Surely, though, the courts could subject both the state's purposes and the fit between means and ends to a more demanding standard of scrutiny.

Justice Scalia himself developed such a standard in *Nollan* and, on a different occasion, arguably endorsed its application in a context similar to *Lucas*.[25] The trouble with standards, though, is that they are inherently manipulable. Whatever the standard, nobody knows in advance how any given case will shake out—the *Nollan* standard, as noted, being no exception. For this reason, Justice Scalia has repeatedly expressed a firm preference for predictable rules over manipulable standards.[26] In *Lucas*, he applied this general jurisprudential preference to the takings context. Instead of a standard, *Lucas* imposes a bright-line test, which holds that the "background principles" of the state's *common law* of nuisance and property constitute a categorical bar to uncompensated (complete) takings.

Scalia considers this principle sufficiently important to state it three times in the course of his opinion. The most satisfying formulation reads as follows:

> Any limitation so severe [as to prohibit all economically beneficial use of land] cannot be newly legislated or decreed (without compensation), but must inhere in the title itself, in the restrictions that background principles of the State's law of

---

24. Lucas, 112 S. Ct. at 2898 n. 12. Justice Scalia's barb is directed against Justice Blackmun's dissent, which endorses a "rational basis" standard of review.

25. Keystone Bituminous Coal, 480 U.S. at 513 (Rehnquist, C.J., joined by J. Scalia, dissenting). Had Justice Scalia followed a "higher scrutiny" approach in Lucas, the Court would have reached a near-identical result, and the majority would likely have spoken with a single voice: Justice Kennedy's concurrence, briefly discussed below in this chapter, looks very much like intermediate scrutiny.

26. The best exposition is Scalia, "The Rule of Law as a Law of Rules," *University of Chicago Law Review*, vol. 56 (1989) p. 1175. For a critical comment on Justice Scalia's jurisprudence in general and its application in Lucas, see Kathleen M. Sullivan, "Foreword: The Justices of Rules and Standards," *Harvard Law Review*, vol. 106 (1992), p. 24.

property and nuisance already place upon land ownership. A law or decree with such an effect must, in other words, do no more than duplicate the result that could have been achieved in the courts—by adjacent landowners (or other uniquely affected persons) under the State's law of private nuisance, or by the State under its complementary power to abate nuisances that affect the public generally, or otherwise.[27]

This common-law solution to the problem of the nuisance exemption may appear to suggest itself, and it is not wholly unprecedented. There is an intimate relation between nuisance law and property titles: in the language of *Lucas*, uses proscribed as noxious under common law are "not part of [the owner's] title to begin with." And, with respect to titles, the Supreme Court has occasionally said that they "are not created by the Constitution. Rather, they are created and their dimensions are defined by existing rules or understandings that stem from an independent source such as state law," including state *common law*.[28]

27. *Lucas*, 112 S. Ct. at 2900. In n. 16 to this passage, J. Scalia elaborates.

> The principal "otherwise" that we have in mind is litigation absolving the State (or private parties) of liability for the destruction of "real and personal property, in cases of actual necessity, to prevent the spreading of a fire" or to forestall other grave threats to the lives and property of others. [case citations omitted]

In addition to the passage quoted in the text, Lucas contains the following formulations:

> Where the State seeks to sustain regulation that deprives land of all economically beneficial use, we think it may resist compensation only if the logically antecedent inquiry into the nature of the owner's estate shows that the proscribed use interests were not part of his title to begin with. (p. 2899)

And again:

> When . . . a regulation that declares "off-limits" all economically productive or beneficial uses of land goes beyond what the relevant background principles would dictate, compensation must be paid to sustain it. (p. 2901)

28. Webb's Fabulous Pharmacies, Inc. v. Beckwith, 449 U.S. 155, 161 (1980) (citing Board of Regents v. Roth, 408 U.S. 564, 577 [1972]) (holding that interest on an interpleader account constitutes property under the "usual and general" and "long established" common-law rule that interest follows the principal). See also Ruckelshaus v. Monsanto, 467 U.S. 896, 1003 (1984) (health, safety, and environmental data cognizable as trade-secret property right under Missouri law; "not necessary that Congress recognize the data at issue here as property in order for the data to be protected by the Takings Clause").

These occasional references, however, only highlight the innovative nature of *Lucas*. Typically, the Supreme Court cast aside state definitions of property as mere technicalities—usually, for the transparent purpose of evading a takings finding.[29] No modern takings case prior to *Lucas* had drawn the connection between the scope of the police power and state *nuisance* law.[30] Moreover, to the limited extent that pre-*Lucas* decisions were predicated on state law, they tended to refer to state law generically, as opposed to state *common* law. So far from protecting private property, this understanding deprived property of constitutional content and left state legislatures free to define property interests in and out of existence.

One could raise a similar objection against the *Lucas* construction: were a state's nuisance law to evolve to the point at which practically all uses of property were presumptively noxious, *Lucas* would protect owners only to the extent that the legislature's application of the relevant precedents is not "objectively unreasonable."[31] As shown more fully in the next section, however, the point is largely academic. The common-law construction made all the difference in the case at hand: the South Carolina Court—which had earlier, and without considering the common-law question, denied compensation—found on remand from the Supreme Court that Lucas was entitled to compensation after all. Moreover, all members of the *Lucas* Court agreed that the common law would pose a significant obstacle to regulatory endeavors. The sharp disagreement among the justices arose over the normative question of whether state legislatures *should* be so tightly constrained.

Property, Factions, and Environmental Values

As noted, the formalistic, common-law construction of *Lucas* embodies a strong preference for clear, predictable rules over manipulable standards.[32] But this preference is not simply a matter of style or aesthetics but ultimately one of substance: it reflects a profound distrust of legislative motives and majoritarian politics.

---

29. Keystone Bituminous Coal, 480 U.S. at 500, 501 (citing Penn Central, 438 U.S. 104 [1978] and Andrus v. Allard, 444 U.S. 51 [1979]).

30. Even the Keystone dissent relied upon state law for the *definition* of the property interest but viewed "the legitimacy of this purpose [*i.e.* nuisance regulation or not [as]] a question of federal, rather than state, law, subject to independent scrutiny by this Court." Keystone Bituminous Coal, 480 U.S. at 512 (Rehnquist, C., dissenting).

31. Lucas, 112 S. Ct. at 2902.

32. See n. 26 in this chapter and Fisher, "The Trouble with *Lucas*," pp. 1408–9.

Justice Blackmun's dissent in *Lucas* clearly recognized this aspect of the majority opinion. Sharply criticizing the majority's resolve to subject legislators to the constraints of judge-made nuisance law, Blackmun asked rhetorically: "Common-law public and private nuisance law is simply a determination whether a particular use causes harm.... If judges in the 18th and 19th centuries can distinguish a harm from a benefit, why not judges in the 20th century, *and if judges can, why not legislators?*"[33]

The *Lucas* majority's rejoinder is that legislators, unlike judges, are subject to political pressures and incentives that tend to eviscerate property rights. We are all eager to subject our neighbors and fellow citizens to controls that do not apply to us and to secure even small benefits for ourselves at great costs, so long as those costs are borne by others. And when we lean on our elected representatives to accomplish such redistributive purposes, legislators will accede to our demands unless constitutional constraints prevent them from doing so. As the Supreme Court put it in an earlier case, legislatures will attempt to force "some people alone to bear burdens which, in all fairness and justice, ought to be borne by the public as a whole."[34] So far from ensuring fairness, an unchecked political process encourages extortion and expropriation. In the language of *Lucas*, if "private property were subject to unbridled, uncompensated qualification under the police power, the natural tendency of human nature [would be] to extend the qualification more and more until at last private property disappear[ed]."[35]

This premise of distrust may not seem especially remarkable. *Lucas* is not the only recent regulatory takings case to indicate that the Supreme Court is no longer willing to sustain a robust presumption of legislative fairness and regularity; in fact, a distrust of legislative motives has become a standard theme of takings law.[36] And one can argue with some justice that the facts of *Lucas* practically compelled a recognition of factionalism. The Beach Front Management Act was not passed by the local jurisdiction, where everyone lost out because the act prohibited not only new construction but also the rebuilding of destroyed structures. Rather, it was enacted by the state legislature, where the beachfront owners were overwhelmed. Then, too, the state repeatedly revised its stated rationales for the act and, in the end, failed

---

33. Lucas, 112 S. Ct. at 2914 (Blackmun, J., dissenting; emphasis added).
34. Armstrong v. United States, 364 U.S. 40, 49 (1960).
35. Lucas, 112 S. Ct. at 2892–93.
36. In addition to Lucas, see, for example, Nollan, 438 U.S. 825 (1987), and Dolan, 114 S. Ct. 2309 (1994).

to come up with any compelling public purpose. The state's *post hoc* rationalizations suggested strongly that it had singled out a few landowners for expropriation.[37] South Carolina subsequently confirmed these fears: when ordered by the state supreme court to pay compensation, the state acquired title to Lucas's lots and quickly resold them for future development. The state spurned a somewhat lower purchase offer by a private preservationist who promised to keep one of the lots undeveloped.[38]

Still, it bears emphasis that the *Lucas* majority chose to stress the dangers of factionalism and expropriation in the context of environmental regulation. In light of its important (stated) public purposes, such regulation should arguably be subject to greater judicial deference than regulations that lend themselves more readily to rank redistribution. *Lucas* shows that environmental rhetoric is no longer enough to avert the Justices' gaze from redistributive motives and consequence. And it is precisely the common-law construction that induces this agnosticism concerning collective values and governmental purposes.

The point emerges most clearly in juxtaposing Justice Scalia's majority opinion with Justice Kennedy's concurrence. Kennedy is by no means oblivious to the dangers of factionalism; in fact, concerns over legislative expropriations of individual landowners account for his decision to vote with the majority.[39] Still, Justice Kennedy voices severe misgivings about Justice Scalia's reasoning. He states the basic disagreement as follows:

> The common law of nuisance is too narrow a confine for the exercise of regulatory power in a complex and interdependent society. . . . The State should not be prevented from enacting new regulatory initiatives in response to changing conditions, and courts must consider all reasonable expecta-

---

37. The state, for example, argued that the prohibition on construction was necessary to limit hurricane damage through flying debris: Lucas, 112 S. Ct. at 2904. The fact that the state did not compel the destruction of existing buildings strongly suggested that its true motivation lay elsewhere.

38. The preservationist offered $315,000 for one of the lots; the state accepted the developer's bid of $392,500 per lot. "[W]hen its own money was on the table, the state was unwilling to forgo $77,500 to preserve one of the lots whose previous value of $600,000 to the owner it had denied was a compensable loss." William A. Fischel, *Regulatory Takings: Law, Economics, and Politics* (Cambridge: Harvard University Press, 1995), p. 61.

39. See, for example, Lucas, 112 S. Ct. at 2904 (Kennedy, J., noting that "the State did not act until after the property had been zoned for individual lot development and most other parcels had been improved, throwing the whole burden of regulation on the remaining lots.").

tions whatever their source. The Takings Clause does not require a static body of state property law; it protects private expectations to ensure private investment. I agree with the Court that nuisance prevention accords with the most common expectations of property owners who face regulation, but I do not believe this can be the sole source of state authority to impose severe restrictions.[40]

Obviously, Justice Kennedy would permit the state to override private expectations more frequently and under different circumstances than would the majority. But the disagreement goes deeper. The Kennedy opinion ties the protection of private expectations to a functional, social purpose: private expectations enjoy protection because they are conducive to investment. That purpose, in turn, enjoys recognition only to the extent that it is commensurate with other, competing public purposes, and in a "complex and interdependent society," the state must regulate beyond the confines of nuisance law to attain those purposes. As broader regulation becomes a more—and, perhaps, the sole—functional response to new problems, the protection of private expectations becomes correspondingly dysfunctional, and the scope of property rights narrows. While Justice Kennedy realizes that legislation *can* be extortionate and redistributive, he also considers legislatures capable of reasoned deliberation and public-regarding action. The judicial determination of whether a given piece of legislation fits into one or the other category should hang not on a yes-or-no answer to the common-law nuisance question but, under the Kennedy approach, on contextual variables—foremost, the *substance* of the regulation. Justice Kennedy cites environmental concerns as particularly compelling rationale for revising common-law doctrines: "Coastal property may present such unique concerns for a fragile land system that the State can go further in regulating its development and use than the common law of nuisance might otherwise permit."[41]

The problem with this argument from complexity is that it reflects a subjective value judgment rather than an objective assessment of probabilities or of a set of facts about the world. Under Justice Kennedy's approach, the state may go beyond nuisance law and take account of complexity and interdependence in protecting the global environment, but must *not* do so in promoting tourism[42]—although the disappearance of open South Carolina shorelines might well affect

---

40. Ibid. at 2903.
41. Ibid.
42. Ibid. at 2904. Promoting tourism was one of the proffered rationales of the Beachfront Management Act.

tourism more directly and severely than it would affect ocean levels. It is not obviously absurd to suggest that environmental regulation, in light of its important public purposes, should be subject to a greater presumption of public-spiritedness and less suspect of rent seeking and naked redistribution than more traditional forms of regulation. But once this preference for ecological over consumptive economic values is allowed to influence legal determinations, property owners are again subject to arbitrary impositions. If the promotion of tourism is insufficient to sustain severe regulation, what, for example, of the preservation of public access to a "unique" environmental treasure? Environmentalists have argued that

> experience of ocean shorefront is an inalienable right of all people. . . . Without access to nature, humans lose perspective on their limited claims on creation, their duties of nurture, and perhaps on the providence of death. Every inducement should be offered to divert citizens from shopping centers and televisions to woods and shores.[43]

Justice Kennedy would probably resist the expropriation of private owners in the name of "inalienable rights" and "inducements" to ecological sensitivity training. But his argument clearly opens the door for such stratagems. Once values reenter the equation—if only under the headings of "complexity" and "uniqueness"—the balancing invariably gravitates toward the values and away from individual rights: what, after all, are Mr. Lucas's profits compared with our spiritual salvation?

Justice Scalia's opinion, in contrast, leaves no room for such balancing; the common-law construction squeezes values entirely out of the takings analysis. The nuisance criterion alone serves as a proxy for distinguishing reasoned deliberation from redistribution: so long as the legislature engages in nuisance abatement, it is presumed to have acted in a reasoned, public-spirited manner; once it goes beyond these confines, the presumption runs the opposite way, regardless of the substance of the legislation. In contrast to Justice Kennedy's concurrence, then, the majority opinion does not permit claims of complexity or ecological uniqueness to trump the rights of private property owners.

Instead, the common-law construction adopts the perspective of the *owner*. This is not a value decision in favor of economic uses, at least not in the same sense in which the preference for environmental

---

43. J. Peter Byrne, "Green Property," *Constitutional Commentary*, vol. 7 (1990), pp. 239, 247–48.

purposes over individual uses is a value judgment: Mr. Lucas is under no obligation to put his lots to their highest and best use but may instead decide to run them as bird sanctuaries. But whatever his intentions, the owner has an interest in predictable rules that create stable expectations and preserve his exclusive control, provided only that he do no harm to others. This is the point of the common law; in this sense, it may be said to favor the owner.

It is true that the common law of nuisance is notoriously messy and arguably just as fraught with subjective judgments as are legislative determinations or, for that matter, judicial balancing tests. On these grounds, the dissenters in *Lucas* observed that "[t]here simply is no reason to believe that new interpretations of the hoary common law nuisance doctrine will be particularly 'objective' or 'value-free'."[44] But this objection misses the point. At the most elementary level, the property owner must operate within the framework of state nuisance law and must adjust his expectations accordingly *in any event*. Hence, the *Lucas* construction does not create new uncertainties but rather affords constitutional protection against—and compensation for—controls and impositions in excess of common-law principles. But there is a second, more fundamental reason why the common law, for all its messiness, tends to be more protective of property rights than unconstrained legislatures or courts.

Private citizens are unlikely to sue their neighbors unless they have been harmed in some tangible way. For this reason alone, private uses that do not measurably and directly affect others will remain undisturbed. An occasional crank or misanthrope may seek to ruin his neighbors over trivialities. But unlike the legislature, he can reach only his neighbors, not the world at large; and in any event, the law is not apt to look kindly on his endeavor. No judge can create new, universally available common-law rights without simultaneously creating corresponding obligations: my newly discovered right to an unobstructed view is your obligation to abstain from a previously lawful construction project. Because of this reciprocity, and because it is well-nigh impossible to know in advance who will lose and who will gain under a new set of rules, it is exceptionally difficult to use the common law for redistributive purposes—in fact, for any kind of purpose other than to distinguish between mine and yours.

This distinction may be difficult, even arbitrary, in particular cases. But the law is unlikely to create rights that would permit private citizens who have suffered no harm to impose conditions on others. The idea of granting every citizen a personal zoning right or free-floating

---

44. Lucas, 112 S. Ct. at 2914 (Blackmun, J., dissenting).

entitlement to a sound ecosystem is absurd; it is tantamount to legalizing aggression by all against all. Under conditions of reciprocity, "keep out" is the only consistent maxim. The common law naturally gravitates toward it and away from all other purposes. In this sense and in the sense of preserving the owner's exclusive control for any nonnoxious use, the common law *is* particularly and uniquely value free and objective—and obnoxious to legislatures bent on redistribution.

A Line in the Sand

In confining state legislatures to the common law, and in its search for bright lines and clear rules, *Lucas* marks a new beginning in takings law. The common-law construction has even been heralded as an overdue return to the original intent of the takings clause and as the solution to the puzzles that have long vexed takings jurisprudence.[45] A decade or so down the road, this hopeful assessment may well turn out to be correct. For the time being, though, a more modest appraisal is warranted.

On its facts, *Lucas* remains a curious case to engineer a return to common-law precepts: as noted, a much less elaborate and more traditional rationale—for example, a heightened standard of scrutiny—would have sufficed to reach the desired result and, arguably, to curb the abuse of environmental regulation for extortionate purposes. And one must concede that the explicit limitations of *Lucas* to complete wipeouts and to land use (as opposed to commercial and property transactions) affect not only the practical reach but also the theoretical plausibility of the decision. The restriction to complete takings is a throwback to the balancing jurisprudence of values and interests;[46] the limitation to land-use cases is wholly unprecedented. In fact, both restrictions are implausible especially if the dangers of factionalism are as pervasive and as central to an understanding of the takings clause as the *Lucas* construction supposes. Notably, the distinction between complete and partial takings practically invites legislatures to take property in incremental steps that fall just short of triggering the compensation requirement.[47] Both restrictions are unsupported by a plausible theory and are introduced for the pragmatic purpose of limiting the enormous sweep of the common-law construction: without its stated limitations, *Lucas* would threaten to "bring down the whole regulatory

---

45. Douglas W. Kmiec, "At Last, the Supreme Court Solves the Takings Puzzle," *Harvard Journal of Law and Public Policy*, vol. 19 (1995), p. 147.
46. Ibid., p. 156.
47. Fisher, "The Trouble with *Lucas*," pp. 1402–3.

apparatus of the modern regulatory state."[48]

The fact that *Lucas* accomplishes its objective of cleaning out a particularly messy area of takings law only at the price of piling up rubble in the rest of the attic raises the question of why the Supreme Court chose to create an elaborate theoretical edifice to reach an altogether unexceptional result of limited practical consequence. The most plausible answer to this question is the interpretation by Professor Sax, mentioned at the beginning of this chapter: *Lucas* can be understood only as a recognition and explicit repudiation of the ecological assumption that the interconnectedness of all things justifies prohibitions on even the most traditional land uses—in the case at hand, the construction of private homes that could not be shown to have any deleterious environmental effects.

This interpretation alone explains the caustic language of *Lucas and* its refusal to decide the case on more conventional grounds *and* its uncompromising rejection of Justice Kennedy's proposed rationale. Above all, it explains the limitations to complete deprivations and to the land-use context, for it is precisely in this context that the ecological paradigm poses the most direct threat to the time-honored castle-and-moat image of private property.[49] The threat is most severe when the paradigm is used to prohibit all viable economic uses of undeveloped land. If such uses were to become a "nuisance," private property could be "pressed into public service";[50] the right to free use would be transformed into an affirmative obligation to preserve private property in its natural state and as a part of a "functioning ecosystem."[51] One can debate whether *Lucas*, on its facts, posed this challenge. But Justice Scalia clearly thought it did—and drew a line in the sand.

---

48. Sax, "Property Rights and the Economy of Nature," p. 1437.
49. Ibid., p. 1451.
50. Lucas, 112 S. Ct. at 2895.
51. Sax, "Property Rights and the Economy of Nature," p. 1438.

3
Standing to Sue

More closely perhaps than any other question of constitutional or administrative law, standing to sue has been interwoven with the emergence of the environment as a salient political issue.¹ Beginning in the late 1960s and throughout the better part of the 1970s, the judiciary dramatically expanded access to the courts and essentially disconnected standing from the idea that a plaintiff must have suffered some tangible harm. Environmental issues and plaintiffs were the driving force and the principal beneficiaries of this development.² Congress ratified the judicial expansion of standing by authorizing any citizen to enforce environmental statutes against other private parties and to sue environmental agencies for failures to perform nondiscretionary statutory duties. The first such citizen suit provision—unprecedented at the time—appeared in the 1970 Clean Air Act; similar provisions were subsequently written into every major environmental statute. This development was sufficiently dramatic to be viewed, even at the time, as a "divorce of judicial review from the core of common law jurisprudence and the recognition of a rather broad public interest in the functioning of government."³

Over the past five years, however, the federal courts have decisively rejected the extension of standing to environmental and other public interest plaintiffs. They have firmly reasserted standing barriers that are tied to concrete, tangible harms. While this shift falls short of a full-fledged resurrection of common-law jurisprudence, it nonetheless

---

1. Bernard Schwartz, *Administrative Law*, 2d ed. (Boston: Little, Brown & Co., 1984), p. 460 ("The expansion of standing ... has made possible the veritable revolutions that have occurred in the environmental law over the past two decades").

2. Important early cases include Scenic Hudson Preservation Conference v. Federal Power Comm'n, 354 F.2d 608 (2d Cir. 1965), cert. denied, 384 U.S. 941 (1966), appeal after remand, 453 F.2d 463 (2d Cir. 1971), cert. denied, 407 U.S. 926 (1972); and Calvert Cliffs Coordinating Comm. v. Atomic Energy Comm'n, 449 F.2d 1109 (D.C. Cir. 1971).

3. Eva H. Hanks, A. Dan Tarlock, and John L. Hanks, *Cases and Materials on Environmental Law and Policy* (St. Paul: West Publishing Co., 1974), p. 214.

signals a decisive rejection of the ecological paradigm and its normative political assumptions.

Ecological Standing

Environmentalism's premise that everything is connected to everything else dictates a virtually boundless view of standing: since everyone is affected by everything that happens in some corner of the ecosystem, courts may not refuse to entertain complaints merely because the plaintiffs are affected only indirectly and to no greater extent than everyone else. Here, as in takings law, the ecological paradigm is at bottom an attempt to leapfrog the notion of tangible harms as a constitutive element of the legal system.[4]

Standing is a *jurisdictional* question: it asks whether a particular plaintiff may bring a particular complaint, but this inquiry is incidental to the larger question of whether the court has the authority to decide the case before it.[5] (This jurisdictional perspective explains, for example, why courts, which must otherwise confine themselves to the issues raised by the litigants, may and arguably must raise standing questions *sua sponte*, without prompting and even over the objections of the parties.) Under Article III of the Constitution, federal courts have the power to decide "cases and controversies" arising under the Constitution—*all*, but *only*, "cases and controversies." Traditionally, these words were taken to mean disputes *over individual rights*—as distinct from ideological gripes, political questions, or abstract questions of the legality of government conduct. This understanding tied judicial review to a common-law conception of individual rights and thought of coercive interference with private autonomy as the paradigm of government conduct. The courts' business vis-à-vis the government was to enforce the fundamental maxim of the common law: keep off.

This perspective, in turn, is closely linked to concerns arising from the constitutional separation of powers. The harness of individual rights prevents the courts from treading on the political branches' constitutional turf; in particular, it ensures that the judicial oversight of

---

4. The discussion in this section largely follows Jeremy A. Rabkin, *Judicial Compulsions: How Public Law Distorts Public Policy* (New York: Basic Books, 1989), pp. 52–63.

5. At the risk of scandalizing legal scholars and practitioners, I use "standing" in its legal sense and as a shorthand expression for closely related jurisdictional doctrines, such as ripeness, mootness, and failure to state a cause of action. Substantially the same analysis applies to all these doctrines, and all of them have followed a common trajectory.

executive authority extends only to the protection of rights. Were judges to go beyond these limits and grant private citizens who have suffered no particular harm a right to litigate the legality of government conduct, the courts would begin to look and act like supervisory review boards and, in effect, exercise the duty and prerogative—committed, in Article II, section 3, of the Constitution, to the executive branch—to take care that the laws are faithfully executed.

This view of the matter is plainly inconsistent with an ecological, value-oriented perspective. Under traditional, harm-based standing rules, litigants who assert the "value" of going about their business without government interference are systematically favored over litigants—including representatives of the public at large—who wish to assert environmental values. At the risk of repeating the obvious, the response that the former litigants have suffered a harm and the latter have not will not satisfy the ecologist: to his mind, the harms are everywhere, and lost profits cannot trump aesthetic or other environmental harms merely because such injuries are less tangible or because they affect everyone at once. From the ecological perspective, the traditional preference for private orderings and autonomy is simply arbitrary.

In response to the obvious question why environmental values are not an equally arbitrary reference point to determine standing, the ecological argument collapses into an assertion of legislative supremacy. When Congress has expressed a collective preference for particular values (such as environmental protection), it is the judiciary's business to give full force and effect to those values; to do so, the courts must grant access to litigants who can be relied on to articulate them. Whether the particular litigants have suffered a particular injury to their own rights is beside the point.

The evisceration of executive authority is an intended consequence of this perspective. Almost by definition, the exercise of executive discretion in the enforcement and implementation of statutory commands cannot be anything except "shirking" or flagrant lawlessness. Consequently, environmentalists perceive no constitutional problem when litigants act in a capacity of private attorneys general: they and the judges who review agency actions at their instigation are simply holding the executive to its constitutional duty.

This constitutional argument dovetails with environmentalism's perspective on the political economy of regulation, which is driven by a fear of agency "capture." Government agencies have a tendency to subvert the public purposes of the law by tailoring their regulatory programs to the needs of the regulated industries. Such capture is produced by a combination of forces. Notably, the costs of environmental regulation are typically more concentrated than its widely dispersed benefits. The regulated parties usually have a lot at stake and are knowledgeable, aggressive, and persistent in pursuing their interests; thus, absent an organized countervailing force, the policy outcomes will be systematically skewed. Traditional standing rules, in this view, exacerbate the tendency toward capture by giving regulated industries a leg up not only in court but also in the regulatory process itself: producers can threaten to sue the agency, whereas participants who represent the public have to make do without this threat—if, indeed, they are permitted to participate in the first place. Again, the analysis compels standing for public-interest litigants: such standing will level the playing field, promote effective public participation, ensure the full enforcement of statutory mandates, and counteract agency capture. In the late 1960s and early 1970s, the courts came to accept this perspective.

Here, as in the takings area, the ground had been prepared. Under New Deal statutes, the courts more or less routinely granted standing to regulated parties and regulatory beneficiaries who alleged a harm to an interest conferred by statute. The 1946 Administrative Procedures Act, which largely codified the extant case law, made judicial review available to anyone "aggrieved *within the meaning of the relevant statute*," regardless of whether those so aggrieved also had a common-law right. Even so defined, though, standing required a harm to a "legal interest," typically in the nature of a pocketbook injury. The parties who obtained standing under this analysis were *interests* whose legislative recognition and disproportionate stake in the regulatory game differentiated them from the world at large.

In 1970, the Supreme Court replaced this loose framework with an "injury-in-fact" test as the basic constitutional standing requirement. ("The 'legal interest' test," Justice William O. Douglas explained, "goes to the merits,"[6] as distinct from the plaintiff's right to bring the case and the court's authority to hear it.) The intended effect of this shift was an enormous expansion of standing especially in environmental litigation. Injury-in-fact still requires some sort of personal harm that sets the plaintiff apart from the world at large; to this extent, the injury-in-fact test may hamper the ecological project, and public-law advocates have consistently criticized the test as a remnant of outdated, "individualistic" common-law conceptions.[7] Until quite recently, though,

---

6. Data Processing Ass'n v. Camp, 397 U.S. 150, 153 (1970). The Supreme Court explicitly recognized " 'aesthetic, conservational and recreational' as well as economic values." Ibid. at 154 (citing Scenic Hudson, 354 F.2d at 616).

7. For example, Steven L. Winter, "The Metaphor of Standing and the Problem of Self-Governance," *Stanford Law Review*, vol. 40 (1988), pp. 1371, 1394–1409; Richard B. Stewart and Cass R. Sunstein, "Public Programs and Private Rights," *Harvard Law Review*, vol. 95 (1982), pp. 1193, 1322.

the point was of little practical consequence. Nominally, there has never been a specialized "law of environmental standing."[8] Environmental plaintiffs, as all other plaintiffs, must allege and later prove that they have suffered an injury that is "palpable" and "direct," that was "caused" by the alleged unlawful conduct, and that is "redressable" by a favorable decision.[9] As a practical matter, however, the courts required little of environmental litigants and stretched "injury-in-fact" to an extent unrecognizable in any other context. In one notorious case, the Supreme Court granted standing to a group of students who challenged an ICC rate-making decision on the grounds that higher railroad freight rates would, through a long chain of events, interfere with their enjoyment of a public park. Environmental standing "shifted from a significant doctrinal barrier to a nettlesome technicality,"[10] and often less than that.

In recent years, however, environmental plaintiffs have lost their preferred status. Twice over the past five years, in *Lujan v. National Wildlife Federation* (1990) and in *Lujan v. Defenders of Wildlife* (1992), the Supreme Court has denied standing to environmental litigants. Although decided on seemingly narrow and technical grounds, both cases signal the demise of the ecological paradigm. Both suggest that the divorce of judicial review from common-law roots and from separation of powers principles may not be irreversible.

National Wildlife Federation and the Demise of Programmatic Litigation

The plaintiffs in *National Wildlife Federation* complained that the Bureau of Land Management had violated the Federal Land Policy and Management Act of 1976 and the National Environmental Policy Act of 1969 (NEPA) in the conduct of its "land withdrawal review program." (Through this program, the BLM determined whether and when to

---

8. The term, however, has been used occasionally by the Supreme Court. For example, Lujan v. Defenders of Wildlife, 112 S. Ct. 2130, 2154 (1992) (Blackmun, J., dissenting).

9. Allen v. Wright, 468 U.S. 737, 751 (1984).

10. William H. Rodgers, *Handbook on Environmental Law* (St. Paul: West Pub. Co., 1977), p. 23. The case mentioned in the text was *SCRAP,* 412 U.S. 669 (1973). Many environmental cases fail to address standing even when it is clearly a problem. Perhaps the earliest example is the then-pathbreaking decision in Calvert Cliffs, 449 F.2d 1109 (1991). Later examples include Robertson v. Methow Valley Citizens Council, 490 U.S. 332 (1989); Japan Whaling Assn. v. American Cetacean Soc., 478 U.S. 221 (1986); and Environmental Defense Fund v. Massey, 986 F.2d 528 (D.C. Cir. 1993).

return federal lands previously withdrawn from any use to the public domain for, among other things, commercial activity such as mining.) Alleging numerous irregularities in the administration of the program, the National Wildlife Federation sought to enjoin future withdrawal revocations and to set aside all revocations already made.

In the posture in which the case reached the Supreme Court, the standing claims rested on the affidavits of two NWF members who claimed recreational and aesthetic use of land "in the vicinity" of two of the 1,250 tracts of land covered by the agency's program. The Supreme Court found these allegations insufficiently specific to support standing. Essentially, the Court held environmental groups to the same pleading and evidentiary requirements that would apply to any other plaintiffs.[11] This ruling signaled a tightening of environmental standing and cast doubt on the continued validity of some prior environmental standing decisions.[12]

Having found that none of the individual plaintiffs had standing, though, the Court went further and launched a sustained (and, by the dissenters' lights, unnecessary) attack on "programmatic" litigation. Justice Scalia's majority opinion rejected at length the court of appeals holding that the plaintiffs could challenge the entire land withdrawal review program under the judicial review provisions of the Administrative Procedures Act (APA). Under these provisions, judicial review is generally available only for final agency actions, as distinct from internal agency deliberations, policies and directives, or preliminary steps in the policy-making process. In the case at hand, the majority found that the land withdrawal review program was "not an 'agency

---

11. Marla E. Mansfield, "Standing and Ripeness Revisited: The Supreme Court's Hypothetical Barriers," *North Dakota Law Review*, vol. 68 (1992), pp. 1, 55 (National Wildlife Federation decision "sent a distinct signal that rules on standing were not to be made more lenient for environmental claims"). See also Daniel Farber, "The Global Environment and the Rehnquist Court," *Trial*, vol. 28 (1992), p. 73 ("*National Wildlife Federation* did not change the legal test for standing [compared with Sierra Club v. Morton, 405 U.S. 727 (1972)] . . . [b]ut it did show that the court was becoming more serious about the test's application").

12. In particular, the opinion strongly suggests, although it does not say outright, that SCRAP, 412 U.S. 669 (1973), is no longer good law. Stephen M. Macfarlane, "*Lujan v. National Wildlife Federation:* Standing, the APA, and the Future of Environmental Litigation," *Albany Law Review*, vol. 54 (1990), p. 866. Even before National Wildlife Federation, however, the Court had described SCRAP as "surely the very outer limit of the law [of standing]": Whitmore v. Arkansas, 495 U.S. 149, 159 (1990).

action' . . . , much less a 'final agency action.' "[13] The Court conceded that a plaintiff who adequately pleads and supports an injury-in-fact, can of course, challenge individual decisions *made pursuant to* the land withdrawal review program, which clearly constitute final agency stations. But even such a plaintiff

> cannot seek *wholesale* improvement of this program by court decree, rather than in the offices of the Department [of the Interior] or in the halls of Congress, where programmatic improvements are normally made. Under the terms of the APA, respondent must direct its attack against some particular "agency action" that causes it harm.[14]

This argument is a thinly veiled reassertion of Article III, injury-in-fact requirements. The underlying notion is that a particular hiker's loss of the enjoyment of a particular park—as a consequence of a determination made under the land withdrawal review program—is a tangible, meaningful "harm," whereas the National Wildlife Federation's concerns over the entire program are not.

The point is of major consequence. If courts are to hold agencies to the thoroughgoing realization of legislative values, they must be able to intervene frequently, comprehensively, and early in the agency decision making process. By the same token, environmental organizations typically seek rulings of broad applicability, both because they must economize on their resources and because only an early intervention will prevent an agency from getting on the wrong track.[15] But the narrow interpretation of "finality" in *National Wildlife Federation* permits only a common-law–like, after-the-fact review of individual, case-specific government actions and therefore thwarts exactly the sort of systematic policy-oriented litigation that has been the environmental movement's most characteristic and arguably most effective legal strategy.[16] In site-specific lawsuits, environmental organizations will almost

---

13. National Wildlife Federation, 497 U.S. at 890. The Court compared the program to "a 'drug interdiction program' of the Drug Enforcement Administration."

14. Ibid. at 891 (emphasis in original).

15. Edward B. Sears, "*National Wildlife Federation:* Environmental Plaintiffs Are Tripped Up on Standing," *Connecticut Law Review,* vol. 24 (1991), pp. 293, 355–59. See also Jonathan Poisner, "Environmental Values and Judicial Review after *Lujan:* Two Critiques of the Separation of Powers Theory of Standing," *Ecology Law Quarterly,* vol. 18 (1991), pp. 335, 366 (effective litigation often requires programmatic review).

16. Cass R. Sunstein, "What's Standing after *Lujan*? Of Citizen Suits, 'Injuries', and Article III," *Michigan Law Review,* vol. 91 (1992), pp. 163, 187–88.

always be able to plead aesthetic or recreational interests with the specificity required under *National Wildlife Federation*.[17] Only and precisely in programmatic lawsuits does the specificity requirement have real bite: such lawsuits are typically directed at programs so general and amorphous, or they are brought at such an early stage, that it becomes impossible to demonstrate a sufficiently concrete and immediate injury.[18]

Justice Scalia's majority opinion in *National Wildlife Federation* acknowledges this point and responds laconically:

> The case-by-case approach that [our interpretation of the APA] requires is understandably frustrating to an organization such as [the National Wildlife Federation], which has as its objective across-the-board protection of our Nation's wildlife and the streams and forests that support it. But this is the traditional, and remains the normal, mode of operation of the courts.[19]

This matter-of-fact reference to the "normal . . . operation of the courts"—again, a reference to Article III concerns—is remarkable especially in light of the factual context of the case. The plaintiffs were challenging a program that, in their estimation at least, was rampant with abuse and irregularities. Convenience and litigation economy aside, they had good reason to seek programmatic review, namely, to prevent countless site-specific decisions from becoming effectively irreversible.[20] The idea that effectual judicial review should be available

---

17. Bradley J. Larson, "Meeting the Requirements of Standing, A Framework for Environmental Interest Groups," *Hamline Law Review*, vol. 14 (1991), pp. 277, 297 ("aesthetic and environmental harms will still be redressable with careful and intelligent lawyering"). Even in site-specific cases, however, the specificity requirements of National Wildlife Federation may occasionally preclude standing. See, for example, NRDC v. Watkins, 954 F.2d 974 (4th Cir. 1992).

18. See, for example, People for the Ethical Treatment of Animals v. H.H.S., 917 F.2d 15 (9th Cir. 1990). In this case, an animal rights group challenged HHS's practice of awarding research grants without first preparing environmental impact statements. Relying squarely on National Wildlife Federation, the Ninth Circuit held that the plaintiffs' general averments to the effect that the funded research activities might interfere with their members' recreational, aesthetic, and other uses of the San Francisco Bay area were insufficient to survive a motion for summary judgment.

19. National Wildlife Federation, 497 U.S. at 894.

20. Bill Hays, "Standing and Environmental Law: Judicial Policy and the Impact of *Lujan v. National Wildlife Federation*," *Kansas Law Review*, vol. 39 (1991) pp. 997, 1033 (noting the possibility that "the opportunity to prevent environ-

before potentially irreversible damage is done has traditionally justified generous grants of standing to environmental watchdog organizations.[21] But *National Wildlife Federation* does not even pay lip service to the special role of environmental beneficiaries in ensuring fidelity to regulatory purposes and public values. The majority opinion acknowledges the "frustration" that comes from the prohibition on programmatic litigation, but what is being frustrated is not our collective aspiration to a clean environment but only an interest group. In the end, *National Wildlife Federation* disconnects standing entirely from public purposes. It treats environmental protection not as a public value but as a private interest.

This break with values and the role of private enforcers as the central reference points of the standing analysis is easily overlooked since *National Wildlife Federation* seems so closely tied to the general review provision of the APA. But it emerges clearly in subsequent appellate cases. In *Conservation Law Foundation of New England v. Reilly*,[22] for example, New England-based environmental organizations sought

---

mental harm is lost when the lands are opened to mining activity through revocation of a withdrawal").

21. Among some appellate judges, the idea retained currency even after National Wildlife Federation. See Foundation on Economic Trends v. Lyng, 943 F.2d 79, 88 (D.C. Cir. 1991) (Buckley, J., diss.):

> NEPA was designed to ensure that "important [environmental] effects will not be overlooked or underestimated only to be discovered after resources have been committed or the die otherwise cast." The need to fully assess potential harm *before* a project is undertaken is a major justification for the broad test courts have laid down for NEPA standing.

(quoting City of Los Angeles v. NHTSA, 912 F.2d 478, 492 [D.C. Cir. 1990] [citation omitted] [emphasis in original]; quoting Robertson v. Methow Valley Citizens Council, 490 U.S. 332, 349 [1989]). To be sure, this statement hangs on the difference between the general review provision of the APA (under which National Wildlife Federation was brought) and NEPA. But for one thing, NEPA challenges are also brought under the APA, NEPA having no separate judicial review provision. For another thing, there is a plausible argument that a more expansive interpretation of APA would have supported a grant of standing in National Wildlife Federation, and at least one appellate court was so persuaded even after National Wildlife Federation: Idaho Conservation League v. Mumma, 956 F.2d 1058, 1516 (9th Cir. 1992) (granting standing to environmental group in programmatic lawsuit on the grounds that "[t]o the extent that [a programmatic] plan pre-determines the future, it represents a concrete injury that plaintiffs must, *at some point*, have standing to challenge" (emphasis added)).

22. 950 F.2d 38 (1st Cir. 1991).

to compel the EPA to evaluate federal waste sites across the country and to place contaminated sites on the National Priority List, as required by the Comprehensive Emergency Response, Compensation, and Liability Act (CERCLA), more commonly known as Superfund. The plaintiffs submitted the affidavits of members who lived or worked near ten identified sites in New England, and no standing problem arose with respect to these sites. The Court, however, denied the requested *nationwide* relief since the plaintiffs had alleged no injury with respect to sites outside New England. This holding illustrates the practical import of *National Wildlife Federation* on programmatic litigation: short of identifying individual plaintiffs near just about every federal waste site in the country, environmental plaintiffs cannot compel a more aggressive implementation of the evaluation and listing process.

This result is particularly noteworthy since *Conservation Law Foundation* was brought not under the APA but under CERCLA's broadly worded citizen suit provision.[23] Even *National Wildlife Federation* granted that its reservations against programmatic judicial review would not apply where Congress invited such review.[24] Arguably, an environmental citizen suit provision extends such an invitation: the principal purpose of such provisions is to compel the prompt and full implementation of environmental programs, and that purpose is easily thwarted if environmental plaintiffs are prevented from obtaining redress when regulatory agencies stall at an early stage of the game. *Conservation Law Foundation*, however, rejects this argument. While agreeing that *National Wildlife Federation* was "not controlling," the First Circuit nonetheless rejected the plaintiffs' contention that CERCLA's citizen suit provision was sufficiently broad to allow anyone to assert the public interest in the administrator's performance of his duties. It did so on the basis of the same concerns over the judiciary's role that lie at the heart of *National Wildlife Federation:* "Absent a showing of concrete injury, a court should not engage in the management of executive functions on such a broad scale [as requested by plaintiffs]."[25]

Other courts have proven equally prepared to apply the general considerations outlined in *National Wildlife Federation* in cases outside

---

23. The citizen suit provision of CERCLA, 42 U.S.C. § 9659(a)(2) (1988), entitles "any person to commence a civil action on his own behalf . . . against the President or any other officer of the United States . . . where there is alleged a failure . . . to perform any act or duty under this chapter . . . which is not discretionary."

24. National Wildlife Federation, 497 U.S. at 1391.

25. Conservation Law Foundation, 950 F.2d at 43.

the APA context.[26] In this light, it seems fair to say that *National Wildlife Federation* "undermines the basis upon which much environmental litigation since 1970 has been premised."[27] The discussion of standing and finality no longer turns on the crucial role of environmental interest groups in ensuring statutory fidelity but, instead, on the limited role of the courts and its incompatibility with programmatic litigation.

National Wildlife Federation did not relate its standing and finality analysis to a formal theory of the separation of powers; in fact, Justice Scalia's opinion did not once mention the Constitution. Still, the decision suggested that separation of powers concerns were central.[28] In *Defenders of Wildlife*, the Supreme Court confirmed this interpretation.

Defenders of Wildlife

Defenders of Wildlife arose over a rule, promulgated by the secretary of the interior, that limited the applicability of certain intergovernmental consultation requirements under section 7 of the Endangered Species Act to actions taken within the United States or on the high seas. Prior to the issuance of this rule, the requirements had been taken to extend also to actions in foreign countries, and the plaintiffs sought to compel the secretary to reinstate the earlier regulation.

As in *National Wildlife Federation*, the plaintiffs' standing claims rested on two affidavits by members of the plaintiff-group. They asserted that they had traveled to Egypt and Sri Lanka and visited eco-

---

26. See for example, Save Ourselves v. U.S. Army Corps of Engineers, 958 F.2d 659 (5th Cir. 1992). The finality argument of National Wildlife Federation has also migrated into NEPA: Foundation on Economic Trends, 943 F.2d 79 (plaintiff-group denied standing to challenge the Department of Agriculture's failure to prepare environmental impact statement on germplasm preservation program); Public Citizen v. U.S. Trade Representative, 5 F.3d 549 (D.C. Cit. 1993). Some courts, however, avoided the thrust of National Wildlife Federation, principally by reading it as narrowly focused on specific allegations of injury. For example, Idaho Conservation League, 956 F.2d 1508 (granting standing to environmental group challenging Forest Service's refusal to make wilderness designations); Sierra Club v. Robertson, 764 F.Supp. 546 (W.D. Ark. 1991) (environmental group granted standing to challenge Forest Service management plan).
27. Macfarlane, *"Lujan v. National Wildlife Federation,"* p. 867.
28. Ibid., p. 866. ("[T]he tone of Justice Scalia's opinion indicates that standing to challenge agency actions in environmental lawsuits brought under the APA should be guided by the same concern for separation of powers principles that the Court has required in other areas of standing jurisprudence"). See also the passages from *National Wildlife Federation* quoted earlier in this chapter.

logically sensitive areas that had subsequently become the sites of large-scale development projects funded by the U.S. government. The plaintiffs further stated that the projects threatened endangered species at the sites and thus deprived them of the opportunity to observe the animals on further visits at some undefined point in the future. The Court held that these affidavits "plainly contain[ed] no facts . . . showing how damage to the species will produce 'imminent injury'" to the individual plaintiffs.[29]

In addition, the plaintiffs tried to support their standing claims with various nexus theories. They asserted an "ecosystem nexus," which would confer standing on any person using any part of a contiguous (albeit expansive) ecosystem adversely affected by an activity funded by the U.S. government; an "animal nexus," which would confer standing on anyone with an interest in studying or seeing endangered animals anywhere on the globe; and a "vocational nexus," which would authorize lawsuits by anyone with a professional interest (such as a zookeeper). The Court rejected these theories out of hand. Although the Court had sometimes extended environmental standing to what Justice Scalia called "the outermost limit of plausibility," the nexus theories advanced in *Defenders of Wildlife* stretched even those limits. *Some* geographical connection, for example, had always been a prerequisite for establishing an injury-in-fact. Here, in contrast, the ecological paradigm appeared shorn of its legalistic fur, and the Court found it far too thin: fairly sneering at the plaintiffs' theories, the majority dismissed the ecosystem nexus as clearly incompatible with *National Wildlife Federation* and the remaining two as "beyond all reason."[30]

Having drawn this line, the Supreme Court went further. In an opinion that marks *Defenders of Wildlife* as "one of the most important standing cases since World War II,"[31] Justice Scalia's opinion for the Court advanced an extended—and, in modern case law, unprecedented—defense of "the doctrine of standing as an essential element of the separation of powers."[32] The explanation of this "essential" connection comes in the context of a categorical rejection of the plaintiffs' claim that they had standing because the secretary's failure to follow

---

29. Defenders of Wildlife, 112 S. Ct. at 2138.
30. Ibid. at 2139.
31. Sunstein, "What's Standing after *Lujan*?" p. 165.
32. Justice (then-Judge) Scalia's law review article by that title, "The Doctrine of Standing as an Essential Element of the Separation of Powers," *Suffolk University Law Review*, vol. 17 (1983), p. 881, presents a more elaborate version of the argument.

the allegedly required consultation procedures had caused them a "procedural injury." An individual can, of course, enforce "procedural rights" and can even do so "without meeting all the normal standards for redressability and immediacy—but only so long "as the procedures in question are designed to protect some threatened concrete interest of his that is the ultimate basis of his standing."[33] Article III standing requirements cannot be satisfied by a "congressional conferral upon *all* persons of an abstract, self-contained, noninstrumental 'right' to have the Executive observe the procedures required by law."[34]

Defenders of Wildlife, however, does not base this holding entirely or even primarily on the case and controversy requirement. Instead, it invokes a separation of powers principle arising elsewhere in the Constitution:

> To permit Congress to convert the undifferentiated public interest in executive officers' compliance with the law into an "individual right" vindicable in the courts is to permit Congress to transfer from the President to the courts the Chief Executive's most important constitutional duty, to 'take Care that the Laws be faithfully executed,' Art. II, §3.[35]

Defenders of Wildlife is not the modern Supreme Court's first standing case to refer to Article II. Notably, in *Allen v. Wright,* the Supreme Court denied standing to a group of taxpayers who challenged a policy of the Internal Revenue Service of granting tax exemptions to racially discriminatory schools. In the course of its opinion, the Court remarked that a grant of standing to taxpayers would turn judges into "virtually continuing monitors of the wisdom and soundness of Executive action"[36] and went on to caution against judicial usurpations of the president's power under the take care clause. As a matter of official doctrine, the same principles applied to statutory standing cases, that is, cases in which the plaintiff's claims rest not on the Constitution but on a congressional enactment: the Supreme Court routinely proclaimed that Congress can waive "prudential" considerations that would ordinarily militate against standing *but not,* of course, the constitutional, Article III minima of injury-in-fact. Judicial practice, though, was another matter. Statutory standing cases prior to *Defenders of Wildlife* clearly belie the persistently reiterated claim that "general-

---

33. Defenders of Wildlife, 112 S. Ct. at 2142 n.7, 2143 n.8.
34. Ibid. at 2143 (emphasis in original).
35. Ibid. at 2145.
36. Allen v. Wright, 468 U.S. 737, 760 (1984) (quoting Laird v. Tatum, 408 U.S. 1, 15 (1972)).

ized grievances" about government conduct are not justiciable.[37] For all practical purposes, the Court assumed that Congress could create any right it wished, the violation of which would then constitute an injury—regardless of whether such rights had common-law analogs and regardless of separation of powers concerns.

Defenders of Wildlife is unprecedented in its clear and unequivocal rejection of this view and in its firm insistence that *Congress* must observe the concrete injury requirement. Justice Scalia's opinion cites several "taxpayer standing" and "generalized grievance" cases, including *Allen*, for the proposition that the constitutional case or controversy requirement precludes relief for a plaintiff who seeks relief on behalf of the public at large. The concession that all these cases arose directly under the Constitution, as opposed to a congressional enactment, is followed by a categorical statement that

> there is absolutely no basis for making the Article III injury turn on the source of the asserted right. Whether the courts were to act on their own, *or at the invitation of Congress*, in ignoring the concrete injury requirement described in our cases, they would be discarding a principle fundamental to the separate and distinct role of the Third Branch.[38]

This statement explicitly repudiates the notion that the courts may and, indeed, must accept a congressional invitation to police the executive at the behest of private attorneys general. It is a direct challenge to the ecological, values-oriented view of the legal world.

Common-Law Analogs?

More, however, needs to be said. It remains true after *Defenders of Wildlife* that the "injury required by Art. III may exist solely by virtue of 'statutes creating legal rights, the invasion of which creates standing.'"[39] The question is how far this principle extends. On this critical point, the majority opinion is not entirely satisfying. It acknowledges

---

37. Defenders of Wildlife, 112 S. Ct. at 2144 ("To be sure, our generalized-grievance cases . . ."). With the exceptions of National Wildlife Federation and Defenders of Wildlife decisions, there does not appear to be a single instance over the past two decades in which the Supreme Court denied standing in a statutory review case. Cases arising under environmental citizen suit provisions typically do not even discuss standing or related justiciability issues. For a rare exception see Gwaltney of Smithfield Ltd. v. Chesapeake Bay Foundation, Inc., 484 U.S. 49 (1987).

38. Defenders of Wildlife, 112 S. Ct. at 2144–45 (emphasis supplied).

39. Ibid. at 2145 (quoting Warth v. Seldin, 422 U.S. 490, 500 [1975]).

that Congress may "elevat[e] to the status of legally cognizable injuries concrete, *de facto* injuries that were previously inadequate in law" and distinguishes such statutory broadening "'from abandoning the requirement that the party seeking review must himself have suffered an injury.'"[40] But if Congress can broaden the categories of injury, why can it *not* permit suits on the basis of, for example, a "procedural injury"? The reply that the injury must be of a "concrete, de facto" type raises only the further question of what precisely *that* might be.

Professor Sunstein has argued critically that Justice Scalia's opinion in *Defenders of Wildlife* tacitly insists "that the requisite injury in fact be defined in common law-like terms."[41] There is this much to Sunstein's complaint: under *Defenders of Wildlife*, the *objects* of regulation—that is, the regulated industries—can ordinarily obtain standing simply by showing that they are objects, whereas "much more is needed" when a "regulatory beneficiary" claims an injury arising from the government's act of regulating, or failure to regulate, someone else.[42] To this extent, the decision ultimately models injury-in-fact on coercive government interference with private freedom. Nonetheless, Sunstein's contention that *Defenders of Wildlife* injects common-law conceptions of harm into the Constitution seems overdrawn, both as a conceptual matter and in practical terms.

In conceding that Congress can elevate to the status of cognizable injuries only concrete, de facto injuries, Justice Scalia mentions two examples from the Supreme Court's standing case law: the injury to an individual's personal interest in living in a racially integrated community and the injury to a company's interest in marketing its product free from competition.[43] To characterize these claims as common-law–type injuries is a stretch. The alleged right to market one's product free from competition certainly seems concrete and de facto, but economic losses incurred as a consequence of competition are also a classic example of an injury the common law considered *damnum absque injuria*, that is, a harm for which there is no legal remedy. The injury to an interest in living in an integrated community, for its part, presupposes a sort of personal property right in a state of affairs whose maintenance requires the cooperation of multiple parties and, more often than not,

---

40. Ibid. at 2145, 2146 (quoting Sierra Club v. Morton, 405 U.S. 727, 738 [1972]).
41. Sunstein, "What's Standing after *Lujan*?" p. 193.
42. Defenders of Wildlife, 112 S. Ct. at 2137.
43. Ibid. at 1145 (citing, respectively, Trafficante v. Metropolitan Life Ins. Co., 409 U.S. 205 [1972], and Harden v. Kentucky Utilities Co., 390 U.S. 1 [1968]).

comprehensive government intervention. Such a construct is obviously far removed from a common-law right that demarcates a private sphere of autonomy or a voluntary, contractual arrangement.

It is, in fact, *so* far removed from the common lawyer's fist in the face that the line-drawing difficulty arises at the other end of the spectrum of harms: the injury to the personal right to racial integration is hard to distinguish from the injury to the interest in species preservation that was alleged in *Defenders of Wildlife*. As Sunstein himself has pointed out, the Supreme Court has routinely granted standing to civil rights plaintiffs (such as Alan Bakke, the most famous reverse discrimination plaintiff) without requiring them to show that they would have obtained the desired job, contract, or admission *but for* the alleged discriminatory practice. For standing purposes, the relevant injury is not the denial of the desired benefit but the denial of equal treatment, or the diminished opportunity to compete on an equal footing. Arguably, though, the plaintiffs in *Defenders of Wildlife* also suffered a diminished opportunity—in their case, the opportunity to observe certain endangered species—as a result of the government's allegedly unlawful procedures. Thus, Sunstein has argued, "[i]f *Bakke* is right on the standing question, it is not so easy to explain why the same approach would have been wrong in [*Defenders of Wildlife*]."[44]

But the Supreme Court has ignored this connection in *Defenders of Wildlife* and thereafter[45]—for the good reason that the conceptual similarity observed by Sunstein obscures profound contextual differences. A civil rights plaintiff's diminished opportunity to compete on an equal footing is, after all, *his* diminished opportunity to compete in a particular, circumscribed market for a particular benefit (typically, of monetary value). In contrast, the diminished opportunity to observe endangered species somewhere on the globe is not so circumscribed. It is indistinguishable from distress at the mere thought of environmental damage, which no more confers standing than a Hawaiian's distress over discriminatory policies in Maine.[46]

Even so, the invocation of Article II, separation of powers princi-

---

44. Sunstein, "What's Standing after *Lujan*?" p. 204 (discussing Univ. of California Board of Regents v. Bakke, 438 U.S. 265 (1978)).

45. In Northeastern Florida Chapter of Associated General Contractors v. City of Jacksonville, 113 S. Ct. 2297 (1993), the Court held that civil rights plaintiffs challenging a municipal set-aside plan need not make a showing above and beyond a diminished opportunity to compete in order to obtain standing. The Court mentioned Defenders of Wildlife only in passing.

46. Such distress does not confer standing. Allen v. Wright, 468 U.S. at 756 (O'Connor, J.).

ples in *Defenders of Wildlife* still has an air of implausibility. Had the plaintiffs in the case purchased airline tickets to Sri Lanka, as opposed to merely averring an intention to visit the place someday, their alleged injury might have been sufficiently imminent to support standing.[47] But why should the ticket purchase alleviate Article II concerns? What does the imminence of the injury (or the lack thereof) have to do with executive authority?

The distinction between concrete, de facto injuries (which, with a little help from Congress, confer standing) and less tangible or more general types of harm (which do not) is often a matter of degree, and may look arbitrary in particular cases. But to leap from this observation to the conclusion that the concrete injury requirement is itself arbitrary is a symptom of acute lawyers' disease. Almost any distinction will look implausible in the marginal cases that become the stuff of litigation, but may still work well over the vast range of ordinary cases—most of which never surface in court *because* the distinction works tolerably well. And as it happens, the concrete injury requirement does a very effective job of separating regulatory *regimes:* it effectively validates interest-group politics, while barring the open-ended pursuit of values.

This boundary is different from the one marked by common-law rights, which protect a perimeter of private autonomy against political usurpation and reduce complex social relations to private orderings. Unlike common-law rights, concrete injuries fence interest-group politics in, not out; for this reason, it is an exaggeration to say that *Defenders of Wildlife* resurrects common-law conceptions. But the requirement of tangible harms does reduce the complexity of an infinitely interrelated world by bringing it back to the more manageable level of interest-group politics.

Defenders of Wildlife creates no standing problems for, say, dairy farmers who claim protection under milk-marketing orders or for welfare recipients who complain about a denial of benefits. Their claims, while lacking clear common-law analogs, are tangible and individualized. Thus, regardless of whether they are viewed as the objects of regulation or as its intended beneficiaries, the clients of the New Deal and of the welfare state can lay claim to legislatively created entitlements. Civil rights claimants are afforded much the same treatment: they too can ordinarily establish the concrete, de facto injury required by *Defenders of Wildlife*. Not so, however, with environmental and

---

47. Defenders of Wildlife, 112 S. Ct. at 2146 (Kennedy, J., conc.), 2153 (Blackmun, J., diss.). Sunstein, "What's Standing after *Lujan*?" p. 213, argues that the link between the concrete injury requirement and Article II is "nonexistent."

health and safety regulation: *its* intended beneficiaries—or, more precisely and perhaps more to the point, the interest groups that purport to litigate on their behalf—often face difficulties in establishing standing.

The point here is not that *Defenders of Wildlife* is hostile toward environmental plaintiffs or concerns per se or in particular. The distinction between cognizable, de facto injuries and generalized grievances corresponds not principally to the regulatory subject matter or to the plaintiffs' substantive concerns but to institutional dynamics: interest-group politics on the one hand, one-dimensional, values-driven regulation on the other. It is true that environmental plaintiffs face greater problems than others in establishing standing under *Defenders of Wildlife*. But this difficulty arises not because the plaintiffs are environmentalists but because environmental citizen-plaintiffs are particularly prone toward the enforcement of statutory values that are disconnected from anyone's particular interests. In contexts where environmental regulation and litigation assume the more circumscribed—if otherwise unedifying—character of interest-group wrangling under administrative management, environmental plaintiffs may satisfy the constitutional minima of standing by showing that they are affected in a tangible way and to a larger extent than the public at large. The environmental users of a particular resource, for example, may obtain standing by demonstrating a geographical nexus to a harm that affects them in particular. And there is no question that labor unions have standing to challenge rule makings by the Occupational Safety and Health Administration—again because their members are affected more directly and tangibly than the general public.

Conversely, *Defenders of Wildlife* casts severe doubt on some heretofore cognizable civil rights injuries that are disconnected from specific individuals or groups. This is true in particular of so-called testers who apply for housing, mortgage loans, or jobs without any intention of actually obtaining these benefits and for the sole purpose of detecting discrimination. Unlike ordinary civil rights plaintiffs, testers (who are typically employed by civil rights groups) act not on their own behalf or even as representatives of racially or otherwise identifiable groups; they act as private attorneys general or as potentially omnipresent trustees and enforcers of egalitarian values. Earlier Supreme Court cases recognized tester standing; *Defenders of Wildlife* bars it precisely because testers act outside the confines of the entitlement politics that is the ordinary mode of civil rights law and litigation.[48]

---

48. In Havens Realty Corp. v. Coleman, 455 U.S. 363 (1982), the Supreme Court sustained tester standing under sec. 804(d) of the Fair Housing Act, 42

Defenders of Wildlife rests on the premise that common-law rights or even legal interests of a tangible, circumscribed sort pose no threat to executive power, whereas the private enforcement of nonnegotiable, disembodied values effectively deprives the executive of its ability and its constitutional prerogative to determine and to pursue political priorities. This premise is contestable, for it is not quite true that traditional, common-law rights of property have no effect on executive power: in fact, such rights are designed to restrict the executive's range of authority. From this observation, one may proceed to the assertion that, for the sake of symmetry and fairness, parties who do not happen to be traditional property owners should be granted analogous rights to assert *their* values against the executive. They should obtain what Professor Sunstein has approvingly called "a kind of property right in a certain state of affairs."[49]

But this analogy collapses under its own weight. Common-law rights mark a mere boundary, outside which there is ample room for executive discretion. Privately enforceable values, in contrast, are inherently claims to exercise executive authority. The property owner seeks to protect and control what is his; the public interest plaintiff seeks to enhance, protect, and control what is by definition not his alone but *everyone's*. Because the citizen plaintiff's purpose and intent is to commandeer common resources, it is extremely misleading to speak of his claims as property rights, even analogously and in a collective sense. The essence of property rights is not a state of affairs but exclusive control; that is why collective actions in property law require either unanimous consent of the parties or democratic governing struc-

---

U.S.C. § 3604(d) and found that this provision conferred a general "right to receive truthful information about housing opportunities" on "anyone." This precedent is a sitting duck. In a post-Defenders of Wildlife decision, Ragin v. Harry Macklowe Real Estate Co., 6 F.3d 898 (2d Cir. 1993), the Second Circuit continued to rely on Havens Realty in granting standing to four black individuals (none of whom were actively seeking to obtain housing) and to a housing rights organization challenging a series of real estate advertisements in the *New York Times* as violative of the Fair Housing Act. The decision, however, is clearly wrong; as several commentators have observed, the result and the reasoning of Havens Realty cannot be squared with Defenders of Wildlife. See Gene R. Nichol, Jr., "Justice Scalia, Standing and Public Law Litigation," *Duke Law Journal*, vol. 42 (1993), pp. 141, 158 n. 125, and Richard J. Pierce, "*Lujan v. Defenders of Wildlife*: Standing as a Judicially Imposed Limit on Legislative Power," *Duke Law Journal*, vol. 42 (1993), pp. 1170, 1179. See also and more generally Michael E. Rosman, "Standing Alone: Standing under the Fair Housing Act," *Missouri Law Review*, vol. 60 (1995), p. 547.

49. Sunstein, "What's Left Standing after *Lujan*?" p. 191.

tures. When it comes to public values, though, there is no exclusivity, and the collectivity is of a kind that permits no unanimous consent: even if we agree on a particular value (such as environmental protection), it is still an open question how we should best promote it. And the only democratic governance that can remotely be relied on—and the only one to which we have consented—is our elected government. For this reason, the "property rights" we have in our common values or in "certain states of affairs" have traditionally been left in the hands of the executive, as circumscribed by law.[50] The public interest actions that are barred by *Defenders of Wildlife* usurp that executive control in a way in which harm-based legal actions, precisely because they are defined and limited by the plaintiffs' harms, do not.

One may still object that the private enforcement of values is not really a *usurpation* of executive power, inasmuch as Congress invited it. One can argue, in other words, that the Constitution implies no conception of executive power that would distinguish the executive from a supreme legislature's "errand boy."[51] On these grounds, Justice Blackmun's dissent in *Defenders of Wildlife* criticized the majority opinion for an "undue solicitude of executive power and a disrespect for Congress *from which that power emanates.*"[52] But this argument does not deny that the concrete injury requirement separates ordinary politics, which allows for executive discretion, from the politics of values, which does not. In fact, Blackmun's argument implicitly concedes the force of the distinction: precisely *because* litigation that is disconnected from tangible harms is so extraordinary and so destructive of executive authority, Justice Blackmun must seek recourse in an equally extraordinary assertion of a legislative supremacy that effectively erases executive power from the Constitution.

Some Connections of Standing and Takings

Defenders of Wildlife and *Lucas* were handed down within three weeks of each other. Both cases produced virtually the same lineup of votes, dissents, and concurrences, and the respective opinions show remark-

---

50. See Harold J. Krent and Ethan G. Schechtman, "Of Citizen Suits and Citizen Sunstein," *Michigan Law Review*, vol. 91 (1993), pp. 1793, 1811–12.
51. The formulation follows Harvey C. Mansfield, Jr., *Taming the Prince* (New York: Free Press, 1989), p. 2. Mansfield is a staunch defender of an executive that is *more* than an errand boy.
52. Defenders of Wildlife, 112 S. Ct. at 2158–59 (Blackmun, J., diss.) (emphasis added). An argument to the same effect is Pierce, "Standing as a Judicially Imposed Limit," esp. pp. 1198–1200.

able substantive congruences. Although they may appear far apart as a matter of ordinary legal taxonomy, the two decisions provide a powerful illustration of the intrinsic connections between takings and standing under the ecological paradigm.

In both cases, the majority confronted and rejected the legal implications of the ecological principle that everything is connected to everything else—the evisceration of property rights on the one hand, limitless standing to sue on the other. In both cases, the majority relied on common-law arguments but pursued them only so far as was necessary to invalidate expansive regulatory arrangements that reflect ecological premises. Faced with the reality that a rigorous understanding and application of common-law conceptions would threaten the entire regulatory apparatus, both cases drew a line—complete deprivations of viable land use in *Lucas*; concrete injuries in *Defenders of Wildlife*.

Even while so limited in scope, Justice Scalia's majority opinions in the two cases drew rejoinders that are based on identical premises and considerations. In *Defenders of Wildlife*, as in *Lucas*, Justice Blackmun submitted a sharply worded dissent; here, as there, he based his disagreement squarely on an assertion of legislative supremacy. In *Defenders of Wildlife*, as in *Lucas*, Justice Kennedy submitted a concurring opinion. He voiced substantial disagreement with the majority's reasoning and the scope of its conclusions and protested that he was "not willing to foreclose the possibility . . . that in different circumstances a nexus theory similar to those proffered here might support a claim of standing."[53] As in *Lucas*, Justice Kennedy based his position on the concern that the needs of a modern society render common-law arrangements dysfunctional:

> As government programs and policies become more complex and far-reaching, we must be sensitive to the articulation of new rights of action that do not have clear analogues in our common-law tradition. Modern litigation has progressed far from the paradigm of Marbury suing Madison to get his commission or Ogden seeking an injunction to halt Gibbons' steamboat operations.[54]

Both for Justice Scalia and for Justice Kennedy, *Lucas* and *Defenders of Wildlife* are two sides of the same coin. For the former, the ecological paradigm poses a threat because of its enormous potential to eviscerate traditional rights at the one end and to create nontraditional ones at the other. Justice Kennedy, in contrast, conjures up an infinitely com-

---

53. Ibid. at 2146 (Kennedy, J., concurring).
54. Ibid.

plex and interdependent world. Everything a holder of traditional (property) rights does may affect everyone else; therefore, private expectations must be revised. Conversely, everyone *is* affected by everything anyone else may wish to do and should therefore be given a stake in the government's decisions concerning third parties, at least so long as Congress grants such rights.[55] Once the notion of tangible, individualized harm has been eliminated from the legal system, the objects of government coercion and its beneficiaries are equal partners in a boundless political process. This is the political-legal perspective of the ecological paradigm. *Lucas* and *Defenders of Wildlife* firmly reject it.

---

55. To be fair, Justice Kennedy cautions that Congress "must at the very least identify the injury it seeks to vindicate and relate the injury to the class of persons entitled to bring suit." Defenders of Wildlife, 112 S. Ct. at 2147 (Kennedy, J., concurring). This constraint, however, seems rather meaningless. Kennedy *also* maintains that "Congress has the power to define injuries and articulate chains of causation that will give rise to a case or controversy where none existed before." Ibid. at 2146–47. This can mean only that courts should defer to such definitions and articulations even when they seem unrealistic and conjectural.

4
Judicial Review of Environmental Regulation

Environmentalism's perspective on the judicial review of agency action is of one piece with its views on standing; the same general considerations apply. The fulcrum of judicial review is the values embodied in environmental statutes, whether in the form of broad statements of intent or in the form of ostensibly ironclad commitments (for example, the Clean Water Act's goal of fishable and swimmable waters by 1983 and the Clean Air Act's objective of clean air that will protect public health "with an adequate margin of safety").[1] The ecological paradigm demands that courts resolve statutory ambiguities in favor of the underlying commitments and that they give force to values even when the operative portions of environmental statutes—that is, the provisions that prescribe how, when, and by what means the EPA is to achieve the legislative goals—indicate, as they often do, that Congress actually intended a somewhat more circumspect and limited pursuit of environmental objectives than would appear from its ringing declarations of intent. Values must trump political bargains and efforts to introduce countervailing considerations, including, foremost, economic costs. Acting on these presumptions, courts have prohibited the consideration of costs in preserving endangered species or in setting air quality and emission standards, among many other examples.[2]

---

1. See, respectively, 33 U.S.C.A. § 1251(a)(2) (1995 West) (Clean Water Act); 42 U.S.C.A. § 7409(b)(1) (1995 West) (Clean Air Act).

2. See Tennessee Valley Authority v. Hill 437 U.S. 153, 184 (1978) (congressional intent in passing the Endangered Species Act was "to halt and reverse the trend towards species extinction whatever the cost"); see also Natural Resources Defense Council v. EPA, 824 F.2d 1146 (D.C. Cir. 1987) (en banc) (remanding hazardous air pollutant regulation because EPA impermissibly considered cost in setting emission standard); American Petroleum Institute v. Costle, 665 F.2d 1176, 1185 (D.C. Cir. 1981), cert. denied, 455 U.S. 1034 (1982) (EPA prohibited from considering cost in setting ambient air quality standard). While the courts typically relied on absolutist or aspirational statutory lan-

As in the area of standing, the values perspective is tied to capture theory. The principal, paradigmatic government failure, in environmentalism's view, is the unwillingness or inability of regulatory agencies to implement and enforce the stringent statutory mandates enacted by Congress. Environmentalism attributes such underregulation or underenforcement chiefly to the political influence of the interest groups on whose economic power legislative intervention was predicated.³ It attempts to address the problem by curtailing, so far as possible, executive discretion. From the environmentalist perspective, discretion merely creates opportunities for bureaucrats and interest groups to frustrate the thoroughgoing realization of public purposes. Much is gained, and nothing of importance is lost, when these opportunities are eliminated.⁴

The expansive, valued-oriented constructions of legislative intent just described are intended to serve precisely this end. A second device to constrain discretion is close judicial scrutiny of an agency's decision-making process. A "hard look" is called for especially when an agency's policies look less aggressive than the underlying statute would seem to permit or when an agency changes its mind and backs off from regulatory commitments: such situations suggest capture, and they pose a grave risk that legislative values are being ignored. Yet a third tool for curbing discretion and preventing capture lies in strengthening the hand of the "underrepresented" interests, such as consumer or environmental groups, in the regulatory process.⁵ The expansion of standing described in chapter 3 was the most important device to facilitate effective participation by public interest groups. In addition, the courts required agencies to follow "hybrid" administra-

---

guage, their "hard look" occasionally discovered environmental values of overriding importance even where Congress had arguably failed to supply them. See, for example, Citizens to Preserve Overton Park v. Volpe, 401 U.S. 402, 416 (1971) (statute that instructed secretary of transportation to balance various factors found to raise park protection to "paramount importance").

3. Cass R. Sunstein, *After the Rights Revolution* (Cambridge: Harvard University Press, 1990), pp. 84–86 (interest groups) and 98–99 (underenforcement and underregulation).

4. Jeremy A. Rabkin, *Judicial Compulsions: How Public Law Distorts Public Policy* (New York: Basic Books, 1989), p. 284 n. 27 (discussing Ralph Nader's proposal to grant any citizen standing to challenge any regulatory decision through civil litigation and noting that Nader implicitly equates executive discretion with tyranny).

5. Richard B. Stewart, "The Reformation of American Administrative Law," *Harvard Law Review*, vol. 88 (1975), p. 1787.

tive procedures—that is, procedures in excess of statutory requirements.[6]

This aggressive posture characterized the judicial review of environmental and health and safety regulation throughout the 1970s. During the 1980s, however, a much more deferential approach gradually won out. As early as 1978, the Supreme Court put an end to the judicial creation of hybrid procedures.[7] (There has been little judicial dispute or academic debate over administrative procedure since that time.) Seven years later, in the watershed case *Chevron v. NRDC*,[8] the Supreme Court held that an agency can follow any "reasonable" interpretation of legislative language so long as Congress has not clearly spoken to the precise question at issue. *Chevron* not only undercut the one-dimensional, values-driven approach to statutory interpretation that characterizes the ecological paradigm; it also created far more room for discretion in regulatory policy making than adherents of that paradigm would countenance.[9] (The received wisdom among administrative lawyers is that *Chevron* cases can be won only under *Chevron I*, that is, the inquiry into whether Congress has decided the precise question at issue: *unless* Congress is found to have settled the issue, the courts' *Chevron II* inquiry as to whether the agency's interpretation of unclear or indeterminate statutory language was "reasonable" will be too deferential to leave plaintiffs much of a chance.)

This shift from ecological values to judicial deference is every bit as important as the reconstruction of harm-based standing barriers described in chapter 3, and it rests on the same normative political assumptions. Justice Scalia, the principal architect of the Supreme Court's modern standing cases, has also been the most forceful and eloquent proponent of the *Chevron* approach. In both areas, his position is defined by identical institutional concern over the protection of executive authority and over the judiciary's lack of competence and democratic au-

---

6. See, for example, Mobil Oil Corp. v. FPC, 483 F.2d 1238 (D.C. Cir. 1973); Natural Resources Defense Council v. Nuclear Regulatory Commission, 547 F.2d 633 (D.C. Cir. 1976).

7. Vermont Yankee Nuclear Power Co. v. National Resources Defense Council, 435 U.S. 519 (1978).

8. Chevron U.S.A., Inc. v. NRDC, 467 U.S. 837 (1985). The "watershed" designation is Kenneth Starr's: Kenneth W. Starr, "Judicial Review in the Post-*Chevron* Era," *Yale Journal of Regulation,* vol. 3 (1986), p. 283.

9. In addition to Chevron and its progeny, see especially Heckler v. Chaney, 470 U.S. 821 (1985), which held that executive enforcement decisions are presumptively unreviewable.

thority to play an aggressive policy-making role.[10]

Admittedly, one can still find decisions that reflect the continued vitality of ecological presumptions,[11] the Supreme Court's 1995 decision in *Babbitt v. Sweet Home Chapter* being a case in point. At issue in the case was a provision of the Endangered Species Act that prohibits anyone from "harming" an endangered species. The Interior Department's extremely broad interpretation of this term had the practical effect of prohibiting any uses of private land that might be an endangered species habitat. Relying on the broad and ambiguous purposes of the act, the Supreme Court sustained this highly questionable interpretation. (In an angry dissent, Justice Scalia criticized the department's interpretation as not even reasonable under *Chevron II*. Commenting on the majority's invocation of the broad purposes of the act, Scalia denounced the "vice of 'simplistically . . . assum[ing] that whatever furthers the statute's primary objective must be the law' " as "the slogan of the enthusiast, not the analytical tool of the arbiter."[12]

But *Sweet Home Chapter* is an exception. Predominantly, the administrative law of the past decade has been shaped by *Chevron's* presumptions against transcendentalism and for executive responsibility and judicial deference, which stand in polar opposition to the doctrines that flow from the ecological paradigm. To be sure, *Chevron* deference may cut not only against but also for environmental values—for example, when an agency aggressively interprets ambiguous language. But where the ecological paradigm comes packaged with capture theory, judicial deference accepts interest-group politics as a basically sound and legitimate way of making policy. And where the ecological paradigm trumpets the legislature's supremacy and calls on courts to give force to statutory aspirations, deference stresses the importance of executive responsibility and reposes trust in the executive branch's ability to manage interest-group conflicts. Accordingly, the *Chevron*

---

10. See Justice Scalia's opinions for the majority in City of Chicago v. Environmental Defense Fund, 114 S. Ct. 1588 (1994); Dept. of Treasury v. Federal Labor Relations Authority, 494 U.S. 922 (1990); and Sullivan v. Everhart, 494 U.S. 83 (1990). See also Antonin Scalia, "Responsibilities of Regulatory Agencies under Environmental Laws," *Houston Law Review*, vol. 24 (1987), p. 24.

11. See, for example, Les v. Reilly, 968 F.2d 985 (9th Cir. 1992), cert. denied, 113 S. Ct. 1361 (1993) (striking down EPA rule allowing "negligible risk" amounts of cancerous pesticides in processed foods).

12. Babbitt v. Sweet Home Chapter of Communities for a Greater Oregon, 115 S. Ct. 2407 (quoting Rodriguez v. United States, 480 U.S. 522, 526 [1987] [footnote omitted]).

perspective has been correctly diagnosed as a repudiation of the view that environmental values are special and entitled to greater judicial protection than the legislative results of ordinary interest-group politics.[13]

From Deference to Substance?

In recent years, a third view has come into play. This outlook borrows both from hard look and from deference but differs from both in certain normative and practical respects. Sometimes called substantive review,[14] this approach is oriented toward ensuring regulatory results that are reasonable in the sense of doing more good than harm. Substantive review reflects the experts' growing disenchantment with the results of environmental regulation. Adherents on and off the bench—chief among the former, Judge Stephen Williams of the influential Court of Appeals for the D.C. Circuit—view environmental regulation as often unreasonable, wildly inefficient, and more than occasionally counterproductive, and they view these flaws as systemic. The analysis borrows from a vast body of public-choice theory on the political economy of regulation.[15] It is organized not around questions of institutional competence—which animate *Chevron* and its progeny—but around the *incentives* of legislators and administrators and especially the dangers of interest-group politics. This much, substantive review shares with the ecological hard look.

Conversely, substantive review shares with *Chevron*-esque deference a skeptical perspective on absolutist environmental aspirations. In fact, it pushes that skepticism much further, in that it tends to view one-dimensional statutory commands as a principal *source* of systemic agency failures. Its paradigm of failure is not capture and underregulation but regulatory excess and tunnel vision, that is, a tendency on the part of government agencies to pursue regulatory commitments

---

13. Robert Glicksman and Christopher H. Schroeder, "EPA and the Courts: Twenty Years of Law and Politics," *Law and Contemporary Problems*, vol. 54, pp. 249, 281–93, and esp. 292–93.

14. Edward W. Warren and Gary E. Marchant, " 'More Good Than Harm': A First Principle for Agencies and Reviewing Courts," *Ecology Law Quarterly*, vol. 20 (1993), p. 379, call this approach "substantive" or "more good than harm review." I prefer the shorter expression.

15. The literature has reached enormous proportions. A good exposition and fair-minded critique of the public choice perspective on statutory interpretation and judicial review is Daniel A. Farber and Philip P. Frickey, *Law and Public Choice: A Critical Introduction* (Chicago: University of Chicago Press, 1991), pp. 88–115.

without regard to competing considerations and counterproductive effects and to the point of negative marginal returns. These failures require not judicial deference but rather a more probing, substantive review that focuses on the *results* of agency regulation.

This style of review is reflected in three important judicial decisions that set aside and remanded, respectively, Corporate Average Fuel Economy (CAFE) standards issued by the National Highway Traffic and Safety Administration, a proposed EPA ban on asbestos products, and the Occupational Safety and Health Administration's (OSHA's) so-called lockout/tagout regulations of energy-generating devices. The three cases hardly amount to a revolution. They have not been widely cited in subsequent case law, and in two of the cases, the practical results were marginal: the EPA abandoned its crusade against asbestos, while both NHTSA and OSHA clung to their positions, and subsequent judicial decisions sustained regulations that were nearly identical to those originally remanded for further consideration.[16] Nonetheless, all three decisions merit close examination. They present a sophisticated critique of the ecological paradigm, and they illustrate the themes that have come to dominate the judicial review of agency action. I explore these issues after a brief summary of the decisions.

In *Competitive Enterprise Institute v. National Highway Traffic Safety Administration (CEI II)*, the D.C. Circuit remanded for further consideration NHTSA's decision to terminate a rule-making proceeding on whether to lower CAFE standards for the 1990 model year from 27.5 miles to 26.5 miles per gallon. The plaintiffs argued that higher fuel efficiency standards invariably lead automakers to downsize cars, that smaller cars are less safe, and that therefore high CAFE standards result in an increased number of highway fatalities. In a majority opinion written by Judge Stephen Williams, the court found that NHTSA had indeed failed to consider this inescapable trade-off. Instead, "with the help of statistical legerdemain, [NHTSA] made conclusory assertions that its decision had no safety cost at all."[17] The court found this reasoning arbitrary and capricious, remanded the case to the agency, and exhorted it to "provide a genuine explanation" for whatever trade-off

---

16. NHTSA's reaffirmation of its original standard was sustained by the D.C. Circuit as reasonable in light of record evidence submitted by the automakers that the standard had no significant effect on the current automobile fleet composition: Competitive Enterprise Institute v. NHTSA, 45 F.2d 481 (D.C. Cir. 1995). OSHA's lockout/tagout rules were sustained in International Union, UAW v. OSHA, 37 F.3d 665 (D.C. Cir. 1994).

17. Competitive Enterprise Institute v. National Highway Traffic Safety Administration, 956 F.2d 321, 324 (D.C. Cir. 1992).

between car safety and fuel efficiency it would ultimately make.

In the second case, *Corrosion Proof Fittings v. EPA*, the Fifth Circuit Court of Appeals vacated and remanded an EPA decision to ban practically all uses of asbestos under the Toxic Substances Control Act. TSCA instructs the EPA administrator to regulate a product if he finds a "reasonable basis to conclude" that its manufacture or use presents an "unreasonable risk of injury" and directs him to regulate "to the extent necessary to protect adequately against such risk using the least burdensome requirements."[18] TSCA lays out a list of regulatory alternatives, ranked in reverse order of their severity. For asbestos, the EPA had chosen the most burdensome of these options, a near-total product ban. (Major uses of asbestos at the time included insulation materials, cements, building materials, fireproof gloves and clothing, and automobile brake linings.) The Fifth Circuit found this course of action patently unreasonable and unsupported by the evidence. "By choosing the harshest remedy given to it under TSCA," the court said, "the EPA assigned to itself the toughest burden in satisfying TSCA's requirement that its alternative be the least burdensome."[19] The agency could not meet this burden. Indeed, the EPA had not even considered more limited regulatory options; "in its zeal to ban any and all asbestos products, [it] basically ignored the cost side"[20] of regulation.

In the third case, *International Union, UAW v. Occupational Safety & Health Administration* (1991), the Court of Appeals for the D.C. Circuit set aside and remanded OSHA's lockout/tagout rules, which mandate certain safety devices for industrial energy-generating equipment. The core of the opinion for the court—like the majority opinion in *CEI II*, written by Judge Williams—is a painstaking interpretation of Section 3(8) of the Occupational Safety and Health Act, which defines an "occupational safety and health standard" as "a standard . . . *reasonably necessary or appropriate* to provide safe or healthful employment and places of employment."[21] The obvious question is what sort of standard is "reasonably necessary or appropriate." In two earlier decisions, commonly known as the *Benzene* and *Cotton Dust* cases,[22] the Supreme Court had ruled that this language required OSHA to make a threshold finding of "significant risk" before the agency could impose regu-

---

18. 15 U.S.C. § 2605(a).
19. Corrosion Proof Fittings v. E.P.A., 947 F.2d 1201, 1216 (5th Cir. 1991).
20. Ibid. at 1223.
21. 29 U.S.C. § 652(8) (1988) (emphasis added).
22. Respectively, Industrial Union Dept., AFL-CIO v. American Petroleum Institute, 448 U.S. 607 (1980), and American Textile Manufactures v. Donovan, 452 U.S. 490 (1981).

lations. *Benzene* and *Cotton Dust*, however, interpreted section 3(8) as applied to *health* standards, which deal with "toxic materials or harmful physical agents." In issuing such standards, OSHA is constrained not only by section 3(8) but also section 6(b)(5). This much more stringent provision instructs the secretary of labor to adopt "the standard which *most adequately* assures, *to the extent feasible*, . . . that no employee will suffer material impairment of health or functional capacity."[23] *Benzene* and *Cotton Dust* took this to mean that once a "significant risk" is found, OSHA must protect employee health up to the point where a standard would pose such economic or technological problems that it would threaten the competitive stability of an industry.

International Union, in contrast, involved not a health but a *safety* standard. OSHA argued that such standards were not subject to the demanding requirements of section 6(b)(5), and could therefore be laxer. Up to this point, the D.C. Circuit agreed. OSHA further argued, however, that it could set *any* safety standard between the feasibility ceiling of section 6(b)(5) and the "reasonably necessary and appropriate" floor of section 3(8), and, on this score, the court took issue with the agency. The D.C. Circuit found that the statute, as construed by OSHA, would come perilously close to an unconstitutional delegation of legislative powers, inasmuch as it would give the agency the "power to roam between the rigor of section 6(b)(5) standards and the laxity of unidentified alternatives."[24] Turning to alternative statutory constructions that would entail a more circumscribed scope of discretion, the court held that the "reasonableness" language of section 3(8) could plausibly be read as permitting—though not requiring—cost-benefit analysis. It remanded the case and instructed OSHA to reconsider certain factual aspects of the lockout/tagout rules in light of the limits of its discretion under section 3(8).

Although social regulatory agencies are not entirely unaccustomed to judicial remands, the rulings in *CEI II*, *Corrosion Proof Fittings*, and *International Union* came as something of a surprise and an embarrassment. CAFE standards and asbestos regulation rank high on the agenda of environmental and consumer groups and enjoy enormous support in Congress and relatively high public visibility.[25] OSHA's

---

23. 29 U.S.C. § 655(b)(5) (1988) (emphasis added).
24. International Union, U.A.W. v. OSHA, 938 F.2d 1310, 1317 (D.C. Cir. 1991).
25. CAFE standards were an issue in the 1992 presidential campaign. See "Campaign '92: Transcript of the Third Presidential Debate," *Washington Post*, October 20, 1992, and Steven Mufson, "Candidates Trade Variety of Charges: Some True, Some Misleading, Others are Simply Obscure," *Washington Post*, October 20, 1992. On the asbestos controversy (which mostly revolved not

lockout/tagout rules, though more esoteric, are of considerable interest to unions in the regulated industries. NHTSA, EPA, and OSHA had each staked much prestige and political capital—not to mention time and money—on an effort to develop and sustain aggressive regulations. Yet each of them lost, and each found its performance criticized severely and in terms that cast doubt on some fundamental assumptions of environmental regulation.

Intent and Reason

The place to begin the analysis is the courts' perspective on congressional intent. *Corrosion Proof Fittings, International Union,* and *CEI II* do not read statutory language literally. They pay close attention to the text, but they read it against a background assumption that Congress desires to achieve reasonable goals by reasonable means. In a formal sense, this perspective is similar to that of the ecological hard-look doctrine, which also read—or purported to read—statutory language in light of a general presumption that "the legislature was made up of reasonable persons pursuing reasonable purposes reasonably."[26] But whereas a hard look mobilized this presumption to urge judicial solicitude of environmental aspirations, substantive review uses it as a point of *departure* from disembodied environmental values. The imputation of reasonableness serves not as a path toward statutory transcendentalism but as a warrant to introduce trade-offs between environmental and other, competing concerns. If legislators are presumed to be reasonable, an absolute commitment to environmental values (all else be damned) must be merely "symbolic"; it cannot possibly mean what it says.[27] Thus, the courts must impute to Congress a willingness to consider trade-offs between environmental and other objectives, at least

---

around the regulations under TSCA but around a separate statute mandating asbestos removal in public buildings), see Josh Barbanel, "The Making of a Crisis," *New York Times,* September 21, 1993; Matthew L. Wald, "Experts Say the Fear of Asbestos Exceeds the Risk in Schools," *New York Times,* September 4, 1993; Philip J. Landrigan, "Asbestos Anxiety," *New York Times,* September 4, 1993; and Randy Kennedy, "Coping with Disruptions and Fears of Asbestos," *New York Times,* August 8, 1993.

26. Henry M. Hart, Jr., and Albert M. Sacks, *The Legal Process: Basic Problems in the Creation and Application of Law* (Westbury, N.Y.: Foundation Press, 1994), p. 1378. Glicksman and Schroeder, "EPA and the Courts," pp. 259–61, identify it as an element of the hard-look doctrine; Warren and Marchant, "More Good than Harm," pp. 405–8, as an element of the new, "substantive" review.

27. See John Dwyer, "The Pathology of Symbolic Legislation," *Ecology Law Quarterly,* vol. 17 (1990), pp. 233, 300–02, 306–8.

so long as the statutory language does not completely foreclose this interpretation.

In operation, the presumption of reasonableness represents a continuum. When the statutory language explicitly allows for or clearly reflects trade-offs, reasonableness is practically indistinguishable from a more conventional analysis that looks toward the plain meaning of the law. At the other end of the spectrum, a judicial insistence that agencies consider costs and other trade-offs even in the face of uncompromising legislative commitments to specific goals will require a robust presumption of reasonableness. On this scale, *Corrosion Proof Fittings* is the most conventional of the three cases. TSCA repeatedly requires a "reasonable basis" for regulation and aims to prevent "unreasonable risk" by means of the "least burdensome requirements." In light of this language, it is hard to escape the conclusion that "Congress did not enact TSCA as a zero-risk statute," and that the EPA's proposed asbestos ban failed "to give adequate weight to statutory language requiring [the agency] to promulgate the least burdensome, reasonable regulation."[28]

International Union is a more ingenious and perhaps more contestable reconstruction. The decision holds that obscure statutory language must be read to provide at least some policy guidance, which, moreover, must be reasonable as a matter of substance. The structure of the argument has a precedent in the Supreme Court's *Benzene* decision, which initiated a "general practice of applying the non-delegation doctrine [that is, the constitutional principle that Congress may not delegate inherently legislative powers] mainly in the form of 'giving narrow constructions to statutory delegations that might otherwise be thought to be unconstitutional.'"[29] Thus, *International Union* and *Benzene* both depart from the hard-look practice of reading even ambiguous statutory mandates in favor of environmental values. Both decisions offer the nondelegation doctrine as a kind of functional substitute for in-depth, hard-look judicial review and for congressional failures to control agency decision making directly and explicitly. But *International Union* goes further.

The judicial willingness to supply narrowing constructions of statutory language that would otherwise violate the constitutional injunction against standardless delegation is based on the idea that "Congress *can* readily articulate some principle by which the beneficent health and safety effects of workplace regulation are to be traded

---

28. Corrosion Proof Fittings, 947 F.2d at 1215.
29. International Union, 938 F.2d at 1316 (citing Mistretta v. United States, 488 U.S. 361, 373 n. 7 [1989]).

off against adverse welfare effects."³⁰ But the fact that Congress did not actually do so poses the question of what trade-off principles can count as reasonable. The response in *Benzene* and *Cotton Dust* ran along the following lines: do as much, OSHA, as you reasonably can without wrecking entire industries (the feasibility requirement), but do not embark on such expensive ventures without some evidence that there is a genuine health risk (the threshold test of significant risk). This framework is probably close to what Congress meant OSHA to do. It assumes, however, that the significance of a risk can be ascertained in isolation from the costs of abating it. This belief is simply mistaken, and it entails bizarre consequences. On the one hand, OSHA is not only free but *compelled* to impose extravagant costs when regulating marginally significant risks; on the other, the agency is prevented from regulating real (though insignificant) risks even when the benefits of doing so would outweigh the costs.³¹ Between flattening the earth and doing nothing, nothing is worth doing.

International Union, in contrast, strongly suggests a cost-benefit approach. As a substantive matter, this holding permits a closer fit between real-world risks and regulatory responses, thus providing a sounder framework of reasoned decision making than *Benzene*.³² But the rationale is further removed from Congress's actual intent, so far as that intent can be ascertained. To be sure, *International Union* said that cost-benefit analysis is only *one* reasonable interpretation of section 3(8). But if there are other interpretations, the court did not hint at what those might be. Moreover, while disavowing any intention of confining the discretion of OSHA to choose among reasonable evalua-

---

30. Ibid. at 1318 (emphasis supplied).
31. The analysis here and below is consistent with Cass R. Sunstein, "Interpreting Statutes in the Regulatory State," *Harvard Law Review*, vol. 103 (1989), pp. 405, 492–93 (arguing that Congress did not explicitly consider trade-offs under OSH Act and that a regulatory regime that requires "significant risk" but precludes cost-benefit analysis is inconsistent). See also Benzene, 448 U.S. at 667–68 (Powell, J., concurring) (criticizing the plurality opinion's significant risk requirement and arguing for proportionality principle).
32. Admittedly, the analysis here abstracts from the statutory differences between the two cases. In the Benzene context of health standards, where OSHA must regulate up to a point just short of endangering the competitive health of an industry, one may want a high significance threshold. Or so, at least, Judge Williams suggested in International Union, 938 F.2d at 1326. But the definition of "feasibility" that compels such drastic regulation is itself the product of Benzene. The larger point is that the International Union decision, unlike Benzene and Cotton Dust, requires systematic comparisons of costs and benefits, and this point is unaffected by the statutory differences.

tion methods, the court nonetheless provided some guidance as to those methods. It criticized, for example, Fifth Circuit decisions that read cost-benefit as a permissible interpretation of section 3(8) for considering only the economic consequences of regulation to the industry, while disregarding costs borne by consumers and workers[33]—a discussion that can be understood only as instructing OSHA to take such costs into account. Similarly, the court observed that Executive Order 12,291, which requires cost-benefit analysis for major rule makings to the extent permitted by law, "may bear on OSHA's authority to promulgate a safety standard whose benefits fail to outweigh its costs."[34]

CEI II did not on its face involve a question of statutory construction. Ironically, though, it may be the most remarkable example of a counterfactual imputation of reasonableness, as exemplified by a consideration of the relevant trade-offs, into a congressional statute. Congress itself wrote the standard of 27.5 miles per gallon into the 1988 Energy Policy and Conservation Act (EPCA). While the statute permits NHTSA to revise the standard upward or downward for each model year in light of certain considerations, one can argue that Congress intended the 27.5 mpg standard as a base line that should carry a presumption of regularity, and that NHTSA's decision to terminate a rulemaking proceeding to change the standard was simply the equivalent of maintaining it. The majority opinion in CEI II, however, is written as if the EPCA had *no* standard, presumptive or otherwise—as if, that is, NHTSA had to face the trade-offs independently and without guidance as to what the congressional base-line expectation might be.[35]

The point is particularly important because EPCA does not even mention automobile safety as a relevant consideration. The act instructs the agency to consider revisions of the CAFE standard in light of four factors, one of which is technological feasibility, including performance. (The others are economic practicability; the effects of other standards, such as airbag requirements, on fuel efficiency; and the need for energy conservation.) One can argue that Congress meant to include safety under performance—but only with some difficulty.[36]

---

33. International Union, 938 F.2d at 1320; with reference to Asbestos Information Association v. OSHA, 727 F.2d 415, 423 (5th Cir. 1984) ("[t]he protection afforded to workers should outweigh the economic consequences to the regulated industry").

34. International Union, 938 F.2d at 1321.

35. Judge Mikva's dissent in the CEI II decision suggests an argument along these lines: 956 F.2d at 327, 329.

36. The issue played a big role in an earlier round of litigation, Competitive Enterprise Institute v. NHTSA, 901 F.2d 107 (D.C. Cir. 1990). The question in this case was whether petitioners, who expressed concern about car safety, fell

The truth is that the trade-off between safety and energy conservation did not loom remotely as large, and probably loomed not at all, in Congress as it came to loom in the agency's and, subsequently, in the D.C. Circuit's mind.

In light of Congress's silence on the safety issue, Judge Williams observed, NHTSA would have had a fair shot at being upheld under the deferential standard of *Chevron* had the agency terminated the rulemaking proceeding on the basis that EPCA did not require it to consider safety at all.[37] But since the agency insisted that safety *was* a relevant consideration, the D.C. Circuit reached the trade-off between safety and energy conservation—ironically, in deference to the agency's construction of the statute. In a less deferential mood, the court then insisted that if an explanation of the trade-off between safety and energy conservation is required at all, the explanation must be serious and reasoned.

Systemic Failure

Once a court has its eyes on trade-offs among competing and conflicting goals (as opposed to the promotion of aspirational statutory objectives), the perspective on agency failure shifts. None of the three decisions just discussed even mentions the dangers of capture and underenforcement; all are concerned with *excessive* regulation. All three opinions paint the agency actions under review, not as bungling or marginal error on a close and difficult question of fact, but as systemic regulatory failure.

One catches more than a whiff of this disposition in the tenor of *Corrosion Proof Fittings* and *CEI II*: both opinions criticize the agencies' conduct in language once reserved for particularly grave instances of agency recalcitrance and foot-dragging. *Corrosion Proof Fittings* denounced the EPA for a "cavalier attitude toward the use of its own data," for making "a mockery of the requirements and of TSCA," and for ignoring the "plain" requirements of TSCA so as "to reach its desired result."[38] The tone of *CEI II* is even sharper; the majority opinion berated NHTSA for "skirt[ing] an obvious conclusion with two specious arguments," for having "fudged the analysis" with the help of "statistical legerdemain," and for "cowering behind bureaucratic

---

within the "zone of interests" protected by EPCA. After a long and not entirely persuasive discussion, then-Judge Ruth Bader Ginsburg eventually concluded that the petitioners did satisfy the zone of interest requirement.
37. CEI II, 956 F.2d at 323.
38. Corrosion Proof Fittings, 947 F.2d at 1219–1230.

mumbo-jumbo"—all to support a mileage standard that, the record showed, "kills people."[39] *International Union* is more measured in tone but no less critical in substance.[40]

Closely linked to regulatory overreach is the tendency of Congress and of regulatory agencies to evade the critical trade-offs that are invariably involved in environmental and health and safety policy. Congress would much rather vote in favor of totally clean air and absolutely pure water than to confront the unpleasant fact—and to explain to the voters—that these things cost money or to face down the interests whose money is at stake. These incentives prompt Congress to write absolutist statutory mandates and sweeping delegations of power into environmental statutes and to leave the difficult trade-offs to the EPA, with the full knowledge that the mandates are illusory. When the agency fails to attain the lofty goals, Congress shouts about special-interest deals; when EPA moves to impose the requirements needed to get from here to there—and thus makes the affected parties squeal—Congress denounces the runaway bureaucracy. It is a win-win game all around for Congress.[41] The same incentives account for Congress's habit of pretending that salvation lies in some technological gimmick just around the corner; environmental statutes abound with "technology-forcing" policies that seek refuge from unpleasant trade-offs in technological fixes.[42]

Since administrative agencies cannot normally overcome the political resistance and resolve the interest-group conflicts that induced Congress to pass the buck in the first place, decisional evasion extends into the rule-making process. Caught between aspirational mandates and political realities, agencies often find themselves unable to support their decisions with a reasoned explanation. This procedural failure gives the reviewing court an opening to review the substance of agency regulations.

Here again, substantive review bears a formal resemblance to a

---

39. CEI II, 956 F.2d at 324–327.

40. For example, 938 F.2d at 1318 ("OSHA's proposed analysis [of the statutory language] would give the executive branch untrammeled power to dictate the vitality and even survival of whatever segments of American business it might choose").

41. For a scathing critique of this "designed-to-fail" regime, see R. Shep Melnick, "Pollution Deadlines and the Coalition for Failure," *The Public Interest*, vol. 75 (spring 1984), p. 123. An equally uncharitable assessment is David Schoenbrod, "Goals Statutes or Rules Statutes: The Case of the Clean Air Act," *UCLA Law Review*, vol. 30 (1983), p. 740.

42. Stephen Breyer, *Breaking the Vicious Circle: Toward Effective Risk Regulation* (Cambridge: Harvard University Press, 1993), p. 48.

hard look: both place a premium on reasoned decision making. But whereas the hard-look doctrine viewed the requirement principally as a means of ensuring fidelity to statutory purposes (as reconstructed by the court),[43] *CEI II, Corrosion Proof Fittings,* and *International Union* use it as a device to reintroduce real-world concerns into statutes that purport to ignore them. The hard-look doctrine ascribed the gap between the legislature's fantastic asking price and the agency's politically negotiated and ill-explained bid to capture and attempted to arbitrage the spread by cajoling the agency into more aggressive regulation. Substantive review, in contrast, views absolutist statutory mandates not as a yardstick but as a source of systemic regulatory failures. Three of these merit a brief discussion: a reliance on technological "fixes" as a means of trade-off avoidance, a tendency to chase marginal risks, and a concurrent tendency to ignore costs, including costs to health and safety.

Technological Fixes. *CEI II* provides a classic example of an agency's reliance on technological innovation as a way to avoid trade-offs. In support of its claim that higher CAFE standards would not lead to net safety losses, NHTSA argued that technological advances and new safety devices had compensated in the past, and would likely compensate in the future, for the safety losses that result from a reduction in vehicle weight and size. Judge Williams's opinion exposed this contention as a conceit: "The appropriate comparison" is not between levels of car safety over time but "between the world with more stringent CAFE standards and the world with less stringent standards."[44] The fact that technological innovation may compensate for the adverse safety effects of vehicle weight reductions means merely that cars would be even safer without those CAFE-induced reductions.

Corrosion Proof Fittings is equally instructive. The EPA argued that the asbestos ban would itself cause the development of low-cost, adequate substitute products. But on this basis, the court observed, the EPA could "ban any product, regardless of whether it has an adequate

---

43. See Motor Vehicle Manufacturers Ass'n v. State Farm Mutual Automobile Ins., 463 U.S. 29 (1983); Bowen v. American Hospital Ass'n, 106 S. Ct. 2101, 2112 (1986) (plurality opinion) (agency duty of reasoned decision making grounded in "Congress's need to vest administrative agencies with ample power to assist in the difficult task of governing a vast and complex nation"). See also Sidney A. Shapiro and Richard E. Levy, "Heightened Scrutiny of the Fourth Branch: Separation of Powers and the Requirement of Adequate Reasons for Agency Decisions," *Duke Law Journal* (1987), pp. 387, 427–28.

44. CEI II, 956 F.2d at 325.

substitute, because inventive companies soon will develop good substitutes."[45] While conceding that "[a]s a general matter . . . a product ban can lead to great innovation,"[46] the court found that the agency had failed to meet the high burden of showing that there were no reasonable alternatives to banning a product for which no substitutes are presently available.[47]

Chasing Marginal Risk at Prohibitive Cost. According to EPA calculations, the regulations under review in *Corrosion Proof Fittings* would have saved 202 or 148 lives, depending on whether the benefits were discounted, at a cost of $450–800 million, depending on the price of substitute products. The more extravagant portions of the regulation would have saved some seven lives over a period of thirteen years at a cost of $200–300 million; the proposed ban on asbestos shingles alone implied a cost of more than $70 million per life saved. The EPA had to devote considerable ingenuity to bring the numbers even within this range. The agency, for example, discounted the benefits of the ban from the time of exposure rather than from the time of the manifestation of an injury—a procedure that makes sense only on the assumption that people are indifferent between contracting cancer now or twenty years down the road.[48]

By any standard, this is a marginal risk; by any standard, the costs are, in the Fifth Circuit's apt description, "astronomical." The court viewed the EPA's estimates of the costs per life saved as proof of the agency's intent to ban asbestos regardless of the actual risk and at *any*

---

45. Corrosion Proof Fittings, 947 F.2d at 1220.
46. Ibid. Perhaps the opinion grants this too readily. Safety bans on sharp-edged knives, door locks, or light bulbs would lead to "great innovation," but the question is whether society is better off as a result. See Bruce A. Ackerman and Richard B. Stewart, "Reforming Environmental Law," *Stanford Law Review*, vol. 37 (1985), pp. 1333, 1359 (arguing that environmental regulation has prompted much useless and unproductive innovation).
47. 15 U.S.C. 2605(c)(1)(C) of TSCA explicitly directs agency to consider availability of substitutes when contemplating a ban. While the EPA argued that the proposed regulations permitted the issuance of waivers to industries if adequate substitutes were not forthcoming, the court responded that "the agency cannot use its waiver provision to argue that the ban of products with no substitutes should be treated the same as the ban of those for which adequate substitutes are available now." Corrosion Proof Fittings, 947 F.2d at 1220.
48. Ibid. at 1218 (criticizing EPA's discount method). See also ibid. at 1218–1219 (criticizing EPA's double counting of the costs of asbestos use and reliance on "unquantified benefits"), and at 1224–1227 (criticizing EPA's failure to include costs of substitutes for asbestos-lined brakes and asbestos-cement pipe).

cost—pointing out, for example, that "the petitioners' brief and our review of EPA caselaw reveals [that] such high costs are rarely, if ever, used to support a safety regulation."[49] Read for all it is worth, *Corrosion Proof Fittings* suggests that health and safety regulations that cost tens of millions of dollars per life saved might *ipso facto* be unreasonable, at least as long as the statutory scheme permits the regulatory agency to take costs into consideration.

Ignoring the Risks of Risk Regulation. All three decisions suggest that agencies must take the unintended and occasionally self-defeating effects of environmental regulation into consideration. Risk regulation involves choice, and "[c]hoice means giving something up."[50] What is being given up in the pursuit of utopian environmental objectives is not only wealth but also, and paradoxically, health and safety benefits.

CEI II involved a straightforward trade-off between energy savings and automobile safety, and the court held that NHTSA had to face and explain that trade-off. In *Corrosion Proof Fittings*, both the trade-offs and the agency's attempt to obscure them were somewhat more subtle. The question was whether the ban on somewhat dangerous asbestos products would lead to an increased use of an even more dangerous product and thus diminish overall safety. On this score, the court found that the EPA had ignored studies (including its own) that showed this indeed to be the case. The most likely substitutes for asbestos-cement pipe, polyvinyl chloride (PVC) and ductile iron pipe, also contain known carcinogens. In fact, in a separate proceeding, the EPA had estimated the cancer risk of PVC at "twenty deaths *per year*, a death rate that stringent controls might be able to reduce to one *per year*, ... *far in excess of the fractions of a life that the asbestos pipe ban may save each year, by the EPA's own calculations.*"[51]

So, too, with the ban on asbestos-lined brakes, which accounted for the bulk of the projected benefits: in its effort to save lives that might be lost if the product remained in use, the EPA ignored credible expert testimony and other evidence showing that the removal of asbestos brakes from the replacement market would result in an increase in highway fatalities that would swamp the safety benefits of the asbestos ban.[52] The EPA's stated reasons for excluding such regulation-in-

---

49. Ibid. at 1223. See also ibid. (suggesting that statutory requirement to consider costs renders regulation that costs $30–40 million per life saved unreasonable).
50. CEI II, 956 F.2d at 322.
51. Corrosion Proof Fittings, 947 F.2d at 1227 (emphasis in the original).
52. Ibid. at 1224 n. 25 (citing testimony of coauthor of EPA-commissioned study).

duced risks were far from persuasive. With respect to toxic substitute products, the agency raised myopia in its own defense: it simply professed more concern about the continued use and exposure to asbestos than about the risk of substitute products. As to the risk of increased highway fatalities, the EPA contended that nonasbestos brakes were less safe because NHTSA had failed to regulate them. Having thus transferred the victims to the books of another agency, the EPA claimed that in light of NHTSA's promise to regulate nonasbestos brakes in the near future, EPA itself was under no obligation to consider the risk of increased highway fatalities in the asbestos proceeding. The Fifth Circuit, however, rejected the EPA's position that risks could simply be ignored, pointing out that the EPA, "eager to douse the dangers of asbestos, . . . inadvertently actually may increase the risk of injury Americans face."[53] The court reminded the EPA that "a death is a death, whether occasioned by asbestos or by a toxic substitute product,"[54] and held that the EPA's failure to consider the risks posed by substitute products rendered the rules unreasonable.

International Union adds another dimension to the problem of unintended consequences. In a separate concurring opinion, Judge Williams observed that it would be a mistake to view the principal trade-off in risk regulation as one of safety benefits versus pure economic costs. Regulation-induced economic losses, Judge Williams explained, might *themselves* produce (statistical) fatalities. Since "larger incomes can produce health by enlarging a person's access to better diet, preventive medical care, safer cars, greater leisure, etc. . . . , there is no basis for a casual assumption that more stringent regulation will always save lives."[55] In other words, a wealthier society is generally a healthier and safer society.[56] Therefore, to the considerable extent that aggressive health and safety regulations impose large compliance costs, reduce aggregate social wealth, and leave less money available for other health-enhancing measures, they may well cost more statistical lives than they save.

The argument that health and safety regulations may, on balance,

---

53. Ibid. at 1221.
54. Ibid.
55. International Union, 938 F.2d at 1326.
56. The best general, nontechnical treatment of the subject is Aaron Wildavsky, *Searching for Safety* (New Brunswick, N.J.: Transaction Books, 1988). For a dramatic illustration, see Indur Goklany, "Richer Is Cleaner: Long-Term Trends in Global Air Quality," in Ron Bailey, ed., *The True State of the Planet* (New York: Free Press, 1995), p. 339 (rising affluence substantially reduces air pollution).

increase risk and cost lives simply by way of imposing economic costs may sound speculative. But although the connections between regulation-induced economic costs and higher risks may be somewhat less direct than the regulatory trade-offs and paradoxes already described (such as the risk of product substitution), they are nonetheless real. Regulatory expenditures may, for example, lead to lower incomes and higher unemployment, which are associated with higher mortality risks.[57] For another example, federal automobile regulations have increased the average price of new cars by over $2,500, thus forcing especially the poor to drive older, smaller, and less safe cars.[58]

As an empirical matter, the connection between increased wealth and better health is most obvious at low levels of income; consider only the case of countries that are too poor to ensure an adequate supply of safe drinking water. But the relation appears quite strong even in societies that are already wealthy, such as the United States. It is estimated, for example, that even at current U.S. income levels a statistical death may be incurred as a result of regulation-induced economic losses of only $50 million, and possibly much less.[59] Many major health and safety regulations are far more expensive. A systematic consideration of regulation-induced risks would reveal that many aggressive regulations—especially those passed in recent years by EPA and OSHA[60]—do more harm than good and are therefore unreasonable under the analysis suggested by Judge Williams's concurrence in *International Union*.

Summary. In the area of regulatory policy, as in the areas of takings and standing, the ecological project is virtually defined by its attempt to jump beyond the notion of tangible harms and toward aspirations and disembodied values. Substantive review, in contrast, encourages a

---

57. See, for example, Ralph L. Keeney, "Mortality Risks Induced by Economic Expenditures," *Risk Analysis,* vol. 10 (1990), pp. 147, 157–58.

58. Warren and Marchant, "More Good than Harm," p. 391 n. 66 and sources cited there.

59. W. Kip Viscusi, "Mortality Effects of Regulatory Costs and Policy Evaluation Criteria," *RAND Journal of Economics,* vol. 25, no. 1 (1994), p. 94. Most experts have come up with lower numbers. For a survey and summary of the findings, see W. Kip Viscusi, "The Dangers of Unbounded Risk Regulation Commitments" (Conference paper; American Enterprise Institute, 1994), pp. 29–31 and table 5.

60. See Office of Management and Budget, *Regulatory Program of the United States Government* (Washington, D.C.: GPO, 1992), pp. 10–12 (listing regulations with average costs in excess of $100 million per death prevented; all but one of these were issued after 1985 by EPA or OSHA).

systematic analysis of the risk–risk trade-offs involved in health and safety regulation; its entire point is to bring the regulatory enterprise back from the pursuit of one-dimensional values to real-world harms and effects. Thus, substantive review does not simply threaten a few particularly ambitious regulations; it poses a pointed, fundamental challenge to environmentalist presumptions.

To repeat: *CEI II, International Union,* and *Corrosion Proof Fittings* do not represent a uniform trend. Moreover, even if the reasonableness approach embodied in the three decisions were to take hold, it remains to be seen how far the courts will be prepared to push it. While the presumption of legislative reasonableness is easily sustained with respect to statutes that call for or at least permit reasonable regulation and cost-benefit analysis, categorical technology, risk, and health standards that leave no room for discretion or interpretation raise more problematic concerns about judicial competence and authority. Even judges who have great confidence in economic reasoning may hesitate to assume an overly aggressive posture when regulatory agencies act under such ironclad—although often unreasonable—legislative mandates.[61]

But these qualifications are much less important than the underlying trend. First, to the extent that substantive review is uncommon, it competes for judicial acceptance not with the ecological hard look of lore but with *Chevron*-style deference and "plain meaning" analysis—which, as noted at the beginning of this chapter, are also inimical to environmentalist presumptions. Second, *CEI II, International Union,* and *Corrosion Proof Fittings* are unusually explicit in criticizing the fundamental assumptions that drove environmental regulation and its judicial review during the environmental era. But the basic assumptions and analytical tools of substantive review can be found in numerous appellate decisions, including many cases that strike a much more conventional tone and a more deferential posture. In sharp contrast to the environmental era, appellate courts now frequently permit or instruct agencies to weigh costs and benefits, at least under statutes that seem to allow such balancing; instruct agencies to choose the least burdensome regulatory option; permit or instruct agencies to ignore trivial risks; and urge them to compare and consider risk-risk trade-offs.[62]

---

61. An interesting example is American Dental Ass'n v. Martin, 984 F.2d 823 (7th Cir. 1993), in which Judges Posner and Easterbrook, two leading law and economics scholars, largely sustained costly OSHA regulations on occupational exposure to AIDS and hepatitis B. In a sharp dissent, Judge Coffey, relying on International Union, urged reversal of OSHA's rules.

62. See Warren and Marchant, "More Good than Harm," pp. 417–28 and cases cited and discussed there.

In substance, then, *CEI II, International Union,* and *Corrosion Proof Fittings* are not exceptions but rather indicators of a broad and profound shift. The labels or the nominal standards of review are much less important than the judges' *substantive* expectations about agency behavior. Substantive review must, in the end, defer to reasonable agency determinations. Conversely, deference must come to an end when an agency clearly oversteps its bounds or botches matters, and the judges' sense of when that is the case depends on extralegal assumptions and intuitions. Substantive review articulates, and gives shape to, the assumption that the basic regulatory failure is excess, not underregulation; that agencies do not get captured nearly as often as they get carried away. Precisely this assumption now guides even the most consistent and adamant defenders of judicial deference.

In *Sweet Home Chapter,* for example—the Endangered Species Act case in which a narrow majority of Supreme Court engineered a surprising and aberrant reassertion of environmental absolutism—Justice Scalia's dissent relied on the ordinary canons of statutory construction in rejecting the Interior Department's and the majority's claim that the act's prohibition against "harming" an endangered species extends to private activities. But the fervent tone and the single-mindedness of the opinion stem from Justice Scalia's conviction that the department's position "imposes unfairness to the point of financial ruin—not just upon the rich, but upon the simplest farmer who finds his land conscripted to national zoological use."[63]

Between a deference that begins and ends with, on the one hand, a robust premise of judicial overreach and unfairness and, on the other hand, a substantive reasonableness review, there is less difference in practice than in theory.[64] Both look to reasonable results, not lofty aspirations; both view judicial review as a check on bureaucratic self-aggrandizement and systemic overreach, not as a contribution to the uncompromising implementation of absolutist environmental values and mandates. Both amount to an explicit and decisive rejection of the ecological paradigm.

---

63. Sweet Home Chapter, 115 S. Ct. at 2421 (Scalia, J., diss.).

64. Even the theoretical differences are easily exaggerated. It is difficult to see, for example, how either prong of the *Chevron* analysis could operate without a basic premise of reasonableness. And, in fact, *Chevron*'s most articulate proponent, Justice Scalia, articulated something close to the "more good than harm" principle of substantive review prior to his appointment to the bench: Antonin Scalia, "Regulation—The First Year," *Regulation,* January/February 1982, p. 19.

5
Functional Rules for a Dysfunctional System

I have argued that judicial decisions on takings, standing, and judicial review signal the demise of ecological public-law conceptions. I have also argued that the trends in these areas are intimately related: just as the ecological paradigm shapes environmentalism's view of takings, standing, and judicial review, so the demise of environmentalist presumptions in these areas follows the same internal logic. At bottom, the ecological premise of the interconnectedness of all things seeks to transcend legal barriers and distinctions that are based on a notion of tangible harm. In making this leap, environmentalism eviscerates property rights; permits lawsuits and control over executive authority by plaintiffs who claim to represent the world at large; and elevates one-dimensional aspirations over sensible results as the lodestar of judicial review. At the same basic level, the demise of environmentalism resurrects harm-based boundaries as a central concern of the law. These boundaries distinguish nuisances from ecological interdependencies; separate tangible, litigable interests from common values that are committed to executive discretion; and orient judicial review toward measurable results, as distinct from the inspiring but often counterproductive pursuit of values.

This chapter turns from the theoretical and ideological implications of legal doctrines toward their practical consequences: it argues that the doctrines of the environmental era were based on fundamental misperceptions of the substantive problems and the political dynamics of environmental regulation. The chapter shows that ecological doctrines routinely produced regulatory excesses and irrational policy results and, moreover, that the return to harm-based legal doctrines promises to curb those excesses and irrationalities.

Although this inquiry is largely of a normative character, it bears on the explanation of *why* the courts came to reject legal doctrines and presumptions that only a decade earlier appeared to have become paradigmatic, if not altogether uncontested. I suggest here, and I argue more explicitly in chapter 6, that the demise of the ecological paradigm is best viewed as a learning process or a response to widely criticized failures of environmental regulation. As the logic of the ecological paradigm unfolded in the law, the courts realized that the doctrines of the environmental era were dysfunctional and that harm-based legal doctrines provide a much needed corrective to the observed regulatory failures.

Although such an interpretation is far from original,[1] it may seem unduly laborious: the demise of the ecological paradigm could be ascribed more obviously and parsimoniously to the increasingly conservative composition of the federal bench. But the weight of the evidence suggests that more than mere politics has been at work.

First, although judicial appointments by Presidents Ronald Reagan and George Bush have clearly influenced the appellate courts' approach to environmental regulation,[2] they appear to have played less of a role than one might be inclined to think.[3] Second, the opinions discussed in the previous chapters uniformly invested far more theoretical effort than would have been necessary to reach the desired "conservative" results, and they show far more intellectual coherence than rank partisanship would produce. Third, and perhaps most important, the courts' changed attitude toward environmental regulation reflects a growing dissatisfaction among experts with the operation and the

---

1. The general argument follows Colin S. Diver, "Sound Governance and Sound Law," *Michigan Law Review*, vol. 89 (1991), pp. 1436, 1445 (arguing that administrative law doctrines can be understood as applications of a theory of regulatory failure).

2. See, for example, Keith Schneider, "Courthouse a Citadel No Longer: U.S. Judges Curb Environmentalists," *New York Times*, March 23, 1992, and Richard J. Pierce, "Two Problems in Administrative Law: Political Polarity on the District of Columbia Circuit and Judicial Deterrence of Agency Rulemaking, *Duke Law Journal* (1988), p. 300 ("[T]he fate of a major agency policy decision reviewed by the D.C. Circuit will vary with the composition of the panel.").

3. A systematic survey of 290 appellate decisions in environmental cases between 1977 and 1990 found that judicial appointments played a real but, on the whole, fairly modest role in explaining pro- or antiregulatory outcomes. William E. Kovacic, "The Reagan Judiciary and Environmental Policy: The Impact of Appointments to the Federal Courts of Appeals," *Environmental Affairs*, vol. 18 (1991), p. 669. On the basis of a more limited survey of environmental cases decided by the important D.C. circuit, the court's chief judge has disputed the charge that the deciding judges' political orientation usually determines the outcome of environmental cases. Patricia H. Wald, "Regulation at Risk: Are Courts Part of the Solution or Most of the Problem?" *Southern California Law Review*, vol. 67 (1994), pp. 621, 645–46.

results of environmental policy. This dissatisfaction is emphatically not a partisan phenomenon but approaches a consensus: virtually no economist or policy analyst of any political persuasion would defend the existing regulatory regime—taken as a whole—as even tolerably efficient and effective.

Experience with the practical operation of ecological doctrines led the courts to conclude that the regulatory failures so roundly criticized by the experts were, in fact, produced or at least exacerbated by those doctrines and that a return to more traditional, harm-based legal doctrines would help to counteract regulatory pathologies. There is far less agreement on these conclusions than there is on the fact of regulatory failures and on their description. But the logic of the ecological paradigm and the empirical evidence support the courts' analysis.

Regulatory Failures

In its breadth and intensity, the experts' sentiment that environmental regulation is fundamentally flawed is matched only by the equally widespread and enthusiastic political support for the objectives of such regulation. It is uncontested, for example, that much regulation is woefully inefficient and needlessly expensive; only the order of magnitude is seriously disputed.[4] There is similarly broad agreement that social regulation has often been counterproductive. Legislative mandates for stringent regulatory controls and standards, for example, often result in underregulation or no regulation at all, since regulatory agencies will refrain from imposing rules that would put entire industries out of business.[5] Similarly, tough technology standards for new facilities often extend the life span of old technologies and thus exacerbate pol-

---

4. For an overview, see Robert W. Hahn and John A. Hird, "The Costs and Benefits of Regulation: Review and Synthesis," *Yale Journal on Regulation*, vol. 8 (1991), p. 233, and Robert W. Hahn and Thomas D. Hopkins, "Regulation/ Deregulation: Looking Backward, Looking Forward," *The American Enterprise*, July/August 1992, p. 70.

5. John Mendeloff, *The Dilemma of Toxic Substance Regulation: How Overregulation Causes Underregulation at OSHA* (Cambridge: MIT Press, 1988), pp. 2–3; John D. Graham, "The Failure of Agency Forcing: The Regulation of Airborne Carcinogens under Section 112 of the Clean Air Act," *Duke Law Journal* (1985), p. 100; William F. Pedersen, "The Future of Federal Solid Waste Regulation," *Columbia Journal of Environmental Law*, vol. 16 (1991), p. 109 (RCRA standards, by virtue of excessive stringency, lead to exclusion of hazardous materials from list of regulated substances); and generally Cass R. Sunstein, "Paradoxes of the Regulatory State," *Chicago Law Review*, vol. 57 (1990), p. 407.

lution.⁶ And there is near-universal agreement that environmental and health and safety regulation suffers from misplaced and confused priorities and from a lack of policy coordination. The EPA itself concedes that vast amounts of public and private resources are spent on minuscule health risks, whereas substantially greater risks go unregulated.⁷

The disheartening experience with inefficient and counterproductive environmental and health and safety regulation, along with the sustained academic criticism of social regulation, has not failed to impress even scholars who see an urgent need for comprehensive environmental regulation and who are generally quite sanguine about government's ability to pursue collective purposes in an efficient and effective manner. Professor Sunstein, for example, a firm supporter of comprehensive environmental regulation and an enthusiastic advocate of its values and aspirations, has nonetheless emerged as a persistent and occasionally sharp critic of the existing regulatory regime. While insisting that environmental regulation has not been an unmitigated disaster, he readily concedes that it often fails to produce sufficient benefits and sometimes proves actually harmful. "The current regulatory framework," Sunstein says, "does not focus national attention on the central issues—the appropriate nature, extent, and level of risk reduction." Among the sources of the observed inefficiencies and policy failures are

> the government's inability to coordinate different programs that cover related aspects of the same problem; Congress's unwillingness to understand that regulatory programs involve complex tradeoffs among competing social goals; interest group "capture" of the regulatory process (an important but overstated phenomenon); and failure on the part of agencies to deal with regulatory obsolescence over time.⁸

---

6. See, for example, Robert Crandall et al., *Regulating the Automobile* (Washington, D.C.: Brookings Institution, 1986), pp. 89, 96–97 (automobile emission standards slowed fleet turnover and may thus have exacerbated mobile source pollution), and Robert Crandall, *Controlling Industrial Pollution* (Washington, D.C.: Brookings, 1983), chap. 7 (new-source standards for stationary sources exacerbated air pollution).

7. For example, Stephen Breyer, *Breaking the Vicious Circle: Toward Effective Risk Regulation* (Cambridge: Harvard University Press, 1993), pp. 19–23 (describing "random agenda selection" and "inconsistency"), and U.S. Environmental Protection Agency, *Reducing Risk: Setting Priorities and Strategies for Environmental Protection* (Washington, D.C.: Government Printing Office, September 1990), p. 13.

8. Cass R. Sunstein, "Administrative Substance," *Duke Law Journal*, (1991), pp. 607, 627. For similar assessments, see Sunstein, *After the Rights Revolution*

An additional, "especially grave," "structural" problem, Sunstein says, is the "creation of poor incentives"—principally, an excessive reliance on counterproductive command-and-control strategies that produce regulatory paradoxes.

It would be misguided to view the judicial doctrines of the environmental era as the sole cause of these regulatory failures or to view the return to more traditional harm-based doctrines as a ready remedy. Environmental policy is intensely legalistic, and the courts play an important role in its formation. But they are only one among several actors on the scene and intervene only sporadically in a continual regulatory enterprise that is shaped not only by legal rules but also, and predominantly, by powerful economic and political incentives. There is no direct line from judicial doctrines to institutional effects or policy outcomes, and legal rules often have unexpected and unintended consequences.

Still, one must ask what kinds of judicial doctrines and practices are *consistent with* or *conducive to* what kinds of regulatory practices and outcomes. From this quasi-judicial vantage point, the ecological paradigm appears as a principal source of the regulatory failures listed by Sunstein. From the same perspective, harm-based doctrines appear as an antidote to regulatory pathologies.

The ecological paradigm abstracts from tangible harms and results, in the belief that such concerns will only compromise environmental values, ignore complexities, and distract from an impartial consideration of the global commons. From this abstraction flow consequences that correspond broadly to Sunstein's list of regulatory failures. If the world is a seamless web, it must be controlled in a comprehensive, centralized fashion, with an eye solely toward the overriding value of ecological survival. This view practically demands a regulatory system built on command and control, since any uncontrolled action would upset the ecological applecart. Given its vast scope and extravagant ambition, such a system should, in turn, be *expected* to produce massive coordination problems and regulatory paradoxes. An unwillingness to understand complex trade-offs is another way of describing regulatory excesses and the one-dimensional nature of environmental aspirations. And regulatory obsolescence—that is, the tendency of government agencies to cling to programs long after they have become unnecessary or even counterproductive—is greatly exac-

---

(Cambridge: Harvard University Press, 1990), esp. pp. 207, 210; Breyer, *Breaking the Vicious Circle*; and Marc K. Landy, Marc J. Roberts, and Stephen R. Thomas, *The Environmental Protection Agency: Asking the Wrong Questions* (New York: Oxford University Press, 1990).

erbated by a legal system that treats the regulatory status quo as the property of public-interest plaintiffs who represent the universe, or claim to do so.

The remainder of this chapter explores these connections between the environmental paradigm and regulatory failures and explains why harm-based doctrines are more conducive to a tolerably efficient system of environmental regulation. The principal argument is that common-law–inspired, harm-based doctrines provide a natural stopping point and a reference point for judges and regulators. For this reason, such doctrines are unlikely to generate the failures that flow naturally from the ecological paradigm.

Property rights provide the clearest illustration. According to the Supreme Court, for example, the Endangered Species Act provides that we must save every last subspecies whatever the cost to society at large and without regard to the burdens imposed on private landowners.[9] To put it gently, the absurd idea that we are required by law to spend the entire national wealth on the preservation of every single rat is not particularly conducive to the sensible allocation of scarce resources even among environmental goods (not to mention competing social values): if everything is infinitely valuable, there can be no priorities.

So long as the costs fall on the public (as they do, for example, when the ESA conflicts with federal construction projects or with logging on public lands), budgetary constraints and political considerations will at least provide some balance. This circumstance increases the government's temptation to accomplish its purposes by expropriating private owners. One can safely assume, though, that absurd policies do not become more rational simply because their costs are less visible. In fact, pressing private land into public service for the preservation of endangered species produces a paradox of its own: such a policy encourages private owners to destroy endangered species or their habitat before their land is integrated, without compensation, into the public ecosystem.[10]

In contrast, the requirement that compensation must be paid for

---

9. Tennessee Valley Authority v. Hill, 437 U.S. 153, 184 (1978); Babbitt v. Sweet Home Chapter of Communities for a Greater Oregon, 115 S.Ct. 2407 (1995).

10. See Brian F. Mannix, "The Origin of Endangered Species and the Descent of Man," *The American Enterprise* November/December 1992, p. 58. See also U.S. Fish and Wildlife Service, 60 *Federal Register*, February 17, 1995, pp. 9507–8 (Endangered Species Act protections for spotted owl have induced overlogging on private lands where no owls are currently present).

nonnoxious uses of private property introduces a sense of balance. Rigorously applied, such a requirement leaves private owners indifferent between a lot with or without an endangered gnatcatcher and thus eliminates the perverse incentive for preemptive destruction. Moreover, the requirement serves as a rough proxy for a collective willingness-to-pay criterion: it forces legislators and regulators to consider the costs of regulatory activities that otherwise look free. This necessity, in turn, encourages regulators to set priorities and orients them toward relatively cost-effective measures. Only an unconstrained regulatory system becomes boundless and inflicts extravagant costs for illusory gains. Thus, property rights provide a powerful illustration of the connections between legal doctrines and regulatory practices and outcomes.

Precisely because the connections are so direct, however, takings law is less instructive from a theoretical perspective than standing or judicial review in exploring the interplay between shifting legal paradigms and regulatory performance. At the same time, and from a practical perspective, the peculiar limitations of the *Lucas* decision to land use and total wipeouts means that the Fifth Amendment prohibition against uncompensated takings will constrain environmental regulators only in rare instances. The courts' shift from ecological presumptions to harm-based doctrines may appear more subtle in the areas of standing and judicial review than the Supreme Court's ostentatiously formalistic majority opinion in *Lucas*, but the paradigm change is more broad-based, decisive, and of greater practical consequence. For these reasons, the discussion of regulatory failures in the remainder of this chapter focuses on standing and judicial review.

Policy Coordination, Trade-offs, and Paradoxes

Coordination problems in environmental policy are legion. Aggressive enforcement programs aimed at protecting one medium (for example, water) prompt regulated industries to shift pollution into other media;[11] smokestack scrubbers mandated under the Clean Air Act generate hazardous sludge, which then has to be exempted from otherwise applicable waste regulations to avoid massive dislocations in the

---

11. See generally National Advisory Council on Environmental Policy and Technology, Technology Innovation and Economics Committee, *Permitting and Compliance Policy: Barriers to U.S. Environmental Technology Innovation* 27, 36–38, 49–50 (October 1990); and U.S. Environmental Protection Agency, *EPA's Clusters: A New Approach for Environmental Management* (Washington, D.C.: Government Printing Office, 1992), pp. 3–4.

utility industry; and so on.[12] The lack of coordination, widely lamented as a principal and serious failure of environmental policy, has produced urgent calls for legal and institutional reform, ranging from a reduction in the number of congressional oversight committees to a more centralized and independent system of expert oversight to a reorganization of the EPA.[13] One must be skeptical about the extent to which these and other proposals can reduce coordination problems. The existing command-and-control system attempts to determine environmental goals and standards for hundreds of substances emitted into land, air, and water and typically seeks to attain its objectives by mandating the "best available technology" for all but the smallest emitters in each and every industry in each and every state. Such regulation is practically designed to produce coordination problems, tunnel vision, and regulatory paradoxes and will do so regardless of its institutional context. The regulatory system has proved largely immune to reform efforts, and among all institutions, the courts are the least equipped to address the deficiencies.

It seems clear, however, that the existing institutional arrangements, from Congress's fragmented committee structure and its piecemeal consideration of environmental problems to the EPA's media-oriented organization, have not helped matters.[14] In this context, the role of the courts merits attention. At any given time, some 80 percent of the EPA's rule-making proceedings are tied up in litigation.[15] This exposure limits the agency's capacity to set coherent priorities and to coordinate policies. With each court reviewing one problem at a time, each limited by the record before it, the judiciary is ill-suited to supply the requisite coordination. In short, when subjected to constant litiga-

---

12. Bruce A. Ackerman and William T. Hassler, *Clean Coal/Dirty Air* (New Haven: Yale University Press, 1981), pp. 29–39. Breyer, *Breaking the Vicious Circle*, p. 22, lists further examples of inconsistent risk policies.

13. See, for example, Richard J. Lazarus, "The Neglected Question of Congressional Oversight of EPA: *Quis Custodiet Ipsos Custodes* (Who Shall Watch the Watchers Themselves)?" *Law and Contemporary Problems*, vol. 54 (1991), p. 232; and Breyer, *Breaking the Vicious Circle*, pp. 59–61.

14. See Breyer, *Breaking the Vicious Circle*, p. 42 (describing congressional committee structure and piecemeal operation as source of ineffective, confused risk regulation), and Lazarus, "The Neglected Question of Congressional Oversight," pp. 226–32.

15. Gary C. Bryner, *Bureaucratic Discretion: Law and Policy in Federal Regulatory Agencies* (New York: Pergamon, 1987), p. 117 (more than 80 percent of the EPA's regulations are challenged in court), and Council on Environmental Quality, *Sixteenth Annual Report*, pp. 2–3 (1985) (85 percent of EPA's regulations result in litigation).

tion, even the most coherent policy agenda will disintegrate.[16] With respect to the style and substance of judicial review, this consideration suggests that courts should not exacerbate the fragmentation of environmental policy and, if possible, play a positive, integrative role. The task lies in identifying judicial review and standing rules that ensure such a result—not always, but over the vast range of cases.

Judicial Review. Viewed from the perspective of its effects on government's ability to pursue a moderately coherent agenda, judicial review should avoid unnecessary intrusion into policy matters and should generate relatively predictable outcomes. The ecological hard look was originally intended to serve precisely this function—principally by wiping out the bureaucratic discretion that was then perceived as the principal source of policy disintegration.[17] But it is difficult even to remember this aspiration to coherence, so intrusive and unpredictable has the hard look proved in practice. There is a logic to this outcome.

Proponents of the hard look disavowed any intention to intrude into the policy-making arena; they claimed no greater role for the judiciary than to see to the full enforcement of the law as laid down by Congress. But this view slights the difference between enforcing law—in the form of tolerably hard-and-fast rules—and vindicating disembodied, transcendental values or aspirations that have no operational content (except, perhaps, to stipulate that enough is never enough). Once "law" takes this form, and once judges purposefully look away from results and toward the abstract values embodied in the law, courts are casting about in a sea of values, with nothing but a vague "standard of review" to guide them. And since a standard is not an operational rule either, a hard look has meant anything from

---

16. There appears to be near-universal agreement on this point. See, for example, Breyer, *Breaking the Vicious Circle,* pp. 57–59; Edward W. Warren and Gary E. Marchant, "More Good than Harm: A First Principle For Agencies and Reviewing Courts," *Ecology Law Quarterly,* vol. 20 (1993), p. 392; Harold H. Bruff, "Coordinating Judicial Review in Administrative Law," *UCLA Law Review,* vol. 39 (1992), p. 1193 (decrying the "pernicious effects" of "[t]he interaction of centralized executive agencies with decentralized reviewing courts"); and Rosemary O'Leary, "The Impact of Federal Court Decisions on the Policies and Administration of the U.S. Environmental Protection Agency," *Administrative Law Review,* vol. 41 (1989), pp. 549, 561–62.

17. See "Environmental Decision-Making: The Agencies versus the Courts," *Natural Resources Lawyer,* vol. 7 (1974), p. 339, 356 (Leventhal, J., remarking that judicial review in environmental cases should "provid[e] a larger perspective, a coordinating perspective"). Judge Leventhal was a principal architect of hard-look review.

extreme deference to painstaking examination of each step in the agency's reasoning process.[18]

The results have been equally unpredictable. Courts have at times set aside acceptable, cost-effective regulation for technical reasons;[19] at other times they have acquiesced in agency failures to consider regulation-induced risks far in excess of the risk reduction attained by the regulation under review.[20] Whether an agency decision is upheld has often more to do with the agency's cleverness in immunizing its regulation against review or with the deciding judges' political predilection than with the good sense of the judgment under review.[21]

In light of this experience, hard-look review came to be viewed as a source of unreasonable agency decisions, inefficiency, confusion, and policy disintegration.[22] The more probing the standard, the more erratic the results will be. Conversely, a more limited judicial role means a larger role for whatever managerial competence and policy coherence the political branches (and the executive in particular) can muster. Precisely this intuition underlies the deferential approach embodied in *Chevron*. The experience with the disintegrating effects of a hard look (combined with its tendency to freeze agencies into obsolete programs, which I address below) has persuaded even scholars who favor aggres-

---

18. Warren and Marchant, "More Good than Harm," p. 398, n. 112–14 and cases cited in ibid.

19. For example, Environmental Defense Fund v. EPA, 598 F.2d 79 (D.C. Cir. 1978).

20. For example, Safe Bldgs. Alliance v. EPA, 846 F.2d 79 (D.C. Cir. 1988), cert. denied sub. nom. National Gypsum Co. v. EPA, 488 U.S. 942 (1988) (sustaining asbestos abatement regulation costing in excess of $70 million per premature death avoided). See the petitioners' attorneys' post-mortem appraisal of this decision, Warren and Marchant, "More Good than Harm," p. 400 n. 122 (EPA's regulation, strongly encouraging the removal of asbestos from schools, was upheld despite evidence that such removal would result in a *net increase* in asbestos levels in such schools; "It is now widely accepted that requiring removal of undamaged asbestos increases airborne asbestos concentrations and occupational exposures, and therefore ends up increasing human exposure and risk.").

21. Warren and Marchant, "More Good than Harm," p. 400 n. 126 and sources cited in notes 1 and 2 of this chapter.

22. Warren and Marchant, "More Good than Harm," p. 400 (unreasonable decisions are the *"inevitable* result when the agency's end product becomes largely irrelevant, and the courts limit themselves to deciding whether the agency's explanation ... was rational.") (emphasis added) (footnote omitted). See also Jerry L. Mashaw and David L. Harfst, "Regulation and Legal Culture: The Case of Motor Vehicle Safety," *Yale Journal on Regulation*, vol. 4 (1987), pp. 257, 276.

sive environmental regulation to defend a more limited judicial role in environmental policy making.[23] In fact, the debate has come almost full circle. Not only has hard-look review ceased to be the orthodox environmentalist position; most advocates of expansive environmental regulation now defend a more deferential judicial posture and militate against substantive review along the lines of *CEI II*, *Corrosion Proof Fittings*, and *International Union*. Their arguments against this style of review parallel Justice Scalia's arguments for judicial deference—respect for agency authority and expertise, and concerns over the judicial usurpation of political functions. One such critic has described *Corrosion Proof Fittings* as

> so lacking in deference to the agency's exercise of expertise and policy judgment, and so full of attempts to impose on the agency the judges' own views of the proper role of regulation in society, that it is virtually indistinguishable from the documents that OMB prepares in connection with its oversight of EPA rulemaking.[24]

It is only fair to observe that the environmentalist about-face on the appropriate standard of review is partly due to the same factors that account for the judiciary's move in the opposite direction: given a comprehensive apparatus of environmental laws and regulations, regulatory agencies that are dedicated to their mission, and a rather more antiregulatory judiciary, environmentalists will naturally opt for judicial deference. But their argument does raise legitimate theoretical and practical concerns. Substantive review applies a strong presumption of reasonableness to congressional enactments, which may be at odds both with an agency's interpretation of a statute and, if sufficiently robust, with actual congressional intent. This type of review also subjects regulation to a fairly searching scrutiny, which might produce problems similar to those that resulted from hard-look review. On balance, though, substantive review seems likely—more likely, at least, than the available alternatives—to contribute to coordinated and coherent environmental regulation.

---

23. See, for example, Pierce, "Two Problems in Administrative Law."

24. Thomas O. McGarity, "Some Thoughts on 'Deossifying' the Rulemaking Process," *Duke Law Journal* (1992), pp. 1385, 1423. For a view close to my own, see Warren and Marchant, "More Good than Harm," pp. 417, 435–36 (characterizing Corrosion Proof Fittings as deferential with respect to technical policy issues). Ultimately, the disagreement is not about the style or standard of judicial review but about the substantive merits of the regulation under review and the normative political assumptions on which Corrosion Proof Fittings is based.

As noted, hard-look review abstracted from regulatory results and focused on the agency's reasoning process itself, with an eye toward ensuring statutory fidelity. Substantive review, in contrast, asks whether the *result* of the agency's deliberations is reasonable. With respect to the steps along the way, the judicial review cases discussed above are quite deferential. *Corrosion Proof Fittings* left technical decisions such as categorization of different types of asbestos to the EPA's discretion. *International Union,* while in effect requiring OSHA to subject safety regulations to cost-benefit analysis, left the agency largely free with respect to the assumption and the modeling of the analysis. And *CEI II* allowed NHTSA to make *whatever* trade-off between energy efficiency and highway safety it chooses and did not even require the agency to express its choice in numerical terms.[25]

Such restraint seems well advised. Apart from the fact that the agency's claims to expertise are strongest with respect to technical and methodological assumptions, the usefulness of reviewing each step of the reasoning process is limited because even an error-free process does not ensure sensible results. For example, regulations are often based on a long series of worst-case assumptions and estimates—each of which might be defensible in isolation but which, when piled on top of each other, produce risk or exposure standards whose benefits stand in no proportion to their costs. Conversely, given the complicated and conjectural subject matter, technical flaws and questionable judgments calls are well-nigh inevitable, without inevitably producing unacceptable decisions. By focusing on the end result, substantive review provides a check on irrational decisions and focuses, if not "national," at least judicial "attention on the central issue—the appropriate nature, extent, and level of risk reduction."[26]

At the same time, the widespread agreement that systemic failures are serious and pervasive justifies a moderately aggressive judicial posture. As noted, the judicial presumption of reasonableness is aimed at what Professor Sunstein calls, felicitously, "Congress's unwillingness to understand that regulatory programs involve complex tradeoffs."[27] By the same token, substantive scrutiny may help counteract policy fragmentation, tunnel vision, and regulatory paradoxes that arise from the design and the institutional incentives of regulatory agencies. *Corrosion Proof Fittings* is perhaps the most instructive example: the court's

---

25. See, respectively, Corrosion Proof Fittings, 947 F.2d at 1224–29; International Union, 938 F.2d at 1319–21; and CEI II, 956 F.2d at 323–24.
26. See Sunstein, "Administrative Substance," p. 627, quoted on page 88 above.
27. Ibid., p. 628.

insistence that the EPA consider the risks of substitute products, as well as its observation that cost estimates of $70 million per life saved are "rarely, if ever, used to support a safety regulation,"[28] focuses squarely on relevant trade-offs and costs, including costs to safety, that the agency chose to ignore. Where the EPA put on blinders, the court reminded the agency of the larger regulatory context.

It remains true that courts have only a vague sense of that context; by institutional design, they cannot serve as so many mini versions of the Office of Management and Budget. It also remains true that a standard of reasonableness is sufficiently vague to produce inconsistent decisions when applied by different courts in different cases. But quite in contrast to the leading judicial proponents of a hard look, the practitioners of substantive review are acutely aware of the limited institutional competence of courts to produce coherent policies. Judges who practice substantive review are keenly aware of the power of political incentives and distrust disembodied legalism; they do not suffer from an illusion that there is a straight line from "efficient" judicial decisions to rational risk management. Witness the following colloquy between Judges Williams and Douglas Ginsburg:

> JUDGE GINSBURG: Do you think, Judge Williams, that the judiciary can act as a check on bureaucratic self-aggrandizement? . . .
>
> JUDGE WILLIAMS: I think the devices open to bureaucracies for evading judicial decisions are almost infinite.
>
> JUDGE GINSBURG: I would suggest as well that the range of decisions open to the agency, which would all qualify as "reasonable" in light of the record, is almost always fairly broad. One has to share the . . . concern that at every turn the agency will choose the point within that range that best serves its own interest—whether it is interested in budget, or prestige, or power, or having an easy time of it. . . .
>
> Judicial review under the APA, with its largely procedural requirements, is a very loose constraint on the agency, for the reason Judge Williams stated.[29]

Both judges are extremely knowledgeable and sophisticated students of regulatory policy; they know inefficient regulation when they see it, which is far more often than they or the rest of us would like. But neither would mistake the D.C. Circuit for an executive review

---

28. Corrosion Proof Fittings, 947 F.2d at 1223.
29. "Debate: Faction and the Environment," *Ecology Law Quarterly*, vol. 21 (1994), pp. 527, 543–44. Full disclosure requires the author to acknowledge his participation in this debate.

board, and neither is bound to think—far less to act on the presumption—that the problems of environmental regulation would be solved if only the regulators were as bright as *they* are.[30] Courts can and should demand results that can be explained in some coherent fashion; on these grounds, Judge Williams set aside and remanded OSHA's lockout/tagout rules in *International Union* and NHTSA's CAFE standards in *CEI II*. But the courts ultimately cannot force agencies to devise rational policies or to abandon irrational policies to which they are politically committed. Probably in recognition of this fact, Judge Williams and Ginsburg upheld, on the second go-around, OSHA and NHTSA regulations that were virtually identical to those Judge Williams had originally remanded for further explanation and consideration.[31] Neither opinion even hints that a stricter standard of scrutiny should apply when agencies return to court with regulations that failed to pass judicial muster once before—a doctrine that was an integral part of the hard-look universe.[32] Where hard look reacted with heightened suspicion, substantive review reacts—so long as the agency can explain its choice the second time around—with a shrug: what else would one expect?

To put the point somewhat differently: as noted at the end of chapter 4, the public-choice perspective and the concern over systemic regulatory excess that underlie substantive review will often color the practical exercise of *Chevron*-style deference. But it is also true that the public-choice adherent's despair over the judiciary's inability to overcome the incentives that drive the regulatory process will often converge on *Chevron*-esque confidence in executive authority and managerial competence: one way or the other, the judges will leave well enough (or bad enough) alone.

It is quite true that the demand for reasonable results—as distinct from extreme deference—may produce errors and inconsistencies: what seems reasonable to one appellate judge may seem outlandish to

---

30. See Stephen F. Williams, "Risk Regulation and Its Hazards," *Michigan Law Review*, vol. 93 (1995), p. 1498 (reviewing now-Justice Stephen Breyer's book, *Breaking the Vicious Circle*, and criticizing Breyer for overestimating the extent to which rational administrators can overcome warped political incentives).

31. See, respectively, International Union, UAW v. OSHA, 37 F.3d 665 (D.C. Cir. 1994) (Williams, J.) and CEI v. NHTSA, 45 F.3d 481 (D.C. Cir. 1995) (Ginsburg, J.).

32. See, for example, Food Marketing Institute v. ICC, 587 F.2d 1285, 1289–90 (D.C. Cir. 1978); Greyhound Corp. v. ICC, 668 F.2d 1354, 1358 (D.C. Cir. 1981).

the next. But one must weigh this danger against the systemic risks of tunnel vision and avoidance of trade-offs. The final regulations under judicial review are the last link in a long chain of efforts, beginning in Congress, to suppress and ignore inescapable trade-offs. Political unwillingness to deal with the inconvenient complexities of environmental regulation produces decisional evasion and rules that are unreasonable in the elementary sense that they cannot be explained. Substantive review as practiced in *CEI II*, *International Union*, and *Corrosion Proof Fittings* is, in the end, a refusal to defer to such evasion and is based on the presumption that a government that makes life-and-death decisions for its citizens owes them "reasonable candor."[33] Such a demand is no more than a minimal requirement of a moderately rational environmental policy and a functional response to widely recognized regulatory failures.

Standing. While the interplay between agency policies and the style and standard of judicial review has been a standard theme of the legal literature, thorough examinations of the connection between standing and administrative politics and policies are rare. Even while legal scholars of a generally liberal, pro-environmentalist mindset have become quite critical of command-and-control regulation, hard-look review, and the usefulness of the capture paradigm, they have failed to subject citizen standing and its effects on the regulatory process to similar scrutiny. In fact, they have fiercely attacked judicial decisions that limit standing for environmental plaintiffs, notably *Defenders of Wildlife*.[34]

In part, the reluctance to subject the effects of citizen standing to critical scrutiny may have to do with a perception that most agency rules have a habit of showing up in court one way or the other; hence,

---

33. CEI II, 956 F.2d at 323 (reviewing court cannot defer to mere decisional evasion; citing SEC v. Chenery Corp., 332 U.S. 194, 196–97 [1947]), 327 ("When the government regulates in a way that prices many of its citizens out of access to large-car safety, it owes them reasonable candor."). See also International Union, 938 F.2d at 1321 ("Where government makes decisions [on life and safety] for others, it may reasonably be expected to make the trade-offs somewhat more explicitly than individuals choosing for themselves.").

34. A particularly startling example of this schizophrenia is Professor Sunstein, who recognizes that "the citizen suit should be seen as part and parcel of a largely unsuccessful system of command-and-control regulation" ("What's Left Standing after *Lujan*? Of Citizen Suits, 'Injuries' and Article III," *Michigan Law Review*, vol. 91 [1992], p. 221)—and then proceeds to advocate cash bounties for private enforcers and "property rights" in clean air (ibid., pp. 233–35).

it seems more profitable to focus on the style of judicial review than on the question of how a case got into court in the first place. In part, the tendency to treat citizen standing as the sacred cow of environmental politics may stem from political commitments and from a fear of losing what scholars perceive as the participatory or balancing effects of citizen standing. Both the inherent logic of environmental litigation and the empirical evidence, however, point toward a tight connection between citizen suits and the regulatory failures—in particular, the disintegration of policy agendas.[35] The resurrection of meaningful, harm-based standing barriers in environmental litigation may well be the single most useful result of the demise of ecological, public-law conceptions.

To be sure, the environmental agenda is being fragmented by litigation from *all* sides, and higher standing barriers do nothing to curb litigation by the objects of environmental regulation, that is, the regulated industries. The institutional consequences of lawsuits by environmental beneficiaries and by regulated industries, however, are asymmetrical. Industry litigation may delay and, in rare instances, derail an agency program. But its disruptive effects do not remotely compare with those of beneficiary litigation. For regulated firms or industries, the reshuffling of agency priorities is at most an incidental byproduct of a lawsuit designed to save the plaintiffs money; for beneficiaries, setting new and different priorities is the principal purpose of litigation. Programmatic lawsuits that reinforce and sustain regulatory commitments, reorient an agency's resources and priorities, and revamp and, occasionally, create entire regulatory programs from whole cloth are virtually *always* brought by beneficiaries.

In addition to their sprawling, boundless nature, citizen suits have a built-in tendency toward transcendentalism. In contrast to industry plaintiffs, who care about the results of regulation *to themselves,* environmental plaintiffs are committed to the enforcement of transcendental values. They may drag in nominal plaintiffs who claim to be particularly aggrieved by the administrative action under review. But the point of this exercise is solely to surmount standing barriers; it has nothing to do with the substance and the purpose of the litigation, which is to enforce values that are *not* reducible to an injury to someone in particular. (In fact, current standing tests preclude citizen standing for parties that are "in it for the money," such as industry plaintiffs

---

35. To the same effect, see Jeremy A. Rabkin, *Judicial Compulsions: How Public Law Distorts Public Policy* (New York: Basic Books, 1989), pp. 72–75. For a somewhat more skeptical view, see Stephen F. Williams, "Fingers in the Pie," *Texas Law Review*, vol. 68 (1990), p. 1303.

that might benefit from more stringent regulation.)[36] But precisely this orientation toward values—as embodied in the law—directs the policy-making process, judicial review included, away from a managerial, results-oriented approach to environmental regulation. Genuine adversarial disputes between the EPA and environmental plaintiffs are rare; in litigation and before, the parties chiefly negotiate about the permissible spread between manageable and politically feasible policies on the one hand and transcendental statutory values on the other. In this manner, the agency is persistently driven to abstract from results and to chase after elusive but strangely reified values, one policy at a time. The resulting tendencies toward tunnel vision and fragmentation are endemic to environmental citizen-suit litigation.

These propensities are exacerbated by the institutional makeup of the environmental movement. The environmental nonprofit market is intensely competitive; environmental groups compete for the same pool of money from foundations, corporations, and, above all, a relatively homogeneous constituency of members and contributors. Competition leads to product differentiation in terms of style and issue selection. Moreover, since members join and contributors contribute for ideological, nonmonetary reasons, the membership incentives offered by environmental groups tend to take the form of shared enthusiasm and uncompromising commitment. This consideration inclines group leaders toward maximalist, nonnegotiable positions.[37]

A good case can be made that competition and product differentiation are no less beneficial in the environmental advocacy market than in any other market. A weaker but still plausible case can be made that the absolutist, zero-risk positions that are characteristic of environmental advocacy ought to be voiced and considered. But one would not want random priorities and absolutist tunnel vision to drive government policies. Unfortunately, citizen suits, each designed to grab a

---

36. Such parties flunk the "zone of interest test." See, for example, Hazardous Waste Treatment Council v. EPA, 861 F.2d 277, 282–85 (D.C. Cir. 1988); and Hazardous Waste Treatment Council v. EPA, 885 F.2d 918 (D.C. Cir. 1989).

37. On such "ideological" incentives, see Mancur Olson, *The Logic of Collective Action* (Cambridge: Harvard University Press, 1965), pp. 12–19, 159–62, and Andrew S. McFarland, *Common Cause, Lobbying in the Public Interest* (Chatham, N.J.: Chatham House, 1984). Efforts to educate members and contributors about the complex trade-offs among environmental objectives and among environmental and other objectives are fraught with peril. See, for example, Raymond Bonner, "Crying Wolf over Elephants," *New York Times Magazine*, February 7, 1993, p. 18 (fundraising concerns prompted environmental groups to advocate ban on ivory trade, despite knowledge that such ban would contribute to decimation of elephant herds).

piece of the environmental agenda and to tighten a particular regulation or program regardless of the consequences on aggregate social risk or the coherence and consistency of agency priorities, have had precisely this effect of translating the dynamics of environmental advocacy into actual policy.

For an instructive example, one need look no further than *Corrosion Proof Fittings*. An intervenor group, the Natural Resources Defense Council (NRDC), strongly supported the EPA's ban on asbestos cement pipe, notwithstanding credible evidence that PVC pipe, the most likely product substitute, posed a far greater health risk. Indeed, in a separate proceeding, the NRDC had sued the EPA for its failure to regulate PVC more stringently.[38]

In a similar fashion, the NRDC advocates tight restrictions on deep-sea waste dumping, with the full knowledge that such restrictions tend to exacerbate the comparatively greater risks posed by alternative disposals, such as landfills, which the NRDC also opposes. NRDC officials may be right in insisting that such take-what-you-can-get strategies are their only plausible political option.[39] But their institutional incentives are an argument *for*, not against, insulating agency priorities against citizen suits; for, not against the contention that restrictions on citizen standing are fully consistent with efforts to improve the coordination of environmental policies and objectives.

Capture and Regulatory Obsolescence

Professor Sunstein's previously cited characterization of agency capture as "an important but overstated phenomenon" probably encapsulates a now widely held view. This shift from the preconceptions of the environmental era, which tended to view capture as ubiquitous and as the principal threat to effective environmental regulation, is both significant and welcome. Nobody would dispute that interest groups command substantial influence in the regulatory process and that such influence often redirects or blunts regulation.[40] But it is a

---

38. The fact was duly noted by Judge Smith: Corrosion Proof Fittings, 947 F.2d at 1227; citing NRDC v. EPA, 824 F.2d 1146, 1148–49 (D.C. Cir. 1987) *(en banc)*.

39. William Booth, "Scientists Propose Experimental Deep-Sea Dumping of Garbage," *New York Times*, January 11, 1991; Tom Kenworthy, "Researchers Chart Impact of Dumping Sludge at Sea," *Washington Post*, November 12, 1992.

40. See, for example, Jerry L. Mashaw and David L. Harfst, "Regulation and Legal Culture: The Case of Motor Vehicle Safety," *Yale Journal on Regulation*, vol. 4 (1987), p. 257 (influence of automobile industry accounts for slow pace of automobile safety regulation); and P. Culhane, *Public Lands Politics: Interest*

great and unwarranted leap from this general observation to the assumption that the critical failure of environmental regulation is underregulation and underenforcement. Nor can one so easily conclude that underregulation, whether measured by a failure to attain aspirational statutory objectives or by some more sophisticated standard, is invariably attributable to capture.

It is an even greater leap to infer that aggressive judicial intervention will provide an effective remedy for a lack of regulatory zeal. In fact, the overwhelming evidence is to the contrary. In a careful empirical study of the implementation of the Clean Air Act, for example, R. Shep Melnick found that the results of intensive judicial intervention were "neither random nor beneficial."[41] He attributed this result specifically to the courts' simplistic view of capture as the critical political dynamic and to their failure to recognize that regulatory agencies typically have complex reasons, including good reasons, for failing to pursue the most aggressive available course of action.[42] Similarly, studies of programmatic lawsuits—the most extreme manifestation of a capture-based conception of the judicial role—have shown that such interventions typically produce little in the way of tangible benefits but distort political priorities and divert critical resources from arguably more important tasks.[43]

Here again, there is a method to the madness; the obsession with the uncompromising implementation of one-dimensional statutes induced the very policy failures that have rightly come to command attention among scholars and judges. So long as capture is the paradigm of agency failure, tunnel vision will tend to look like commendable commitment, trade-offs will look like cave-ins, and regulatory paradoxes will invite further, even more convoluted interventions. The judicial commitment to assist in the enforcement of environmental values was not the only source of the paradoxical institutional effects of the capture paradigm. But it did exacerbate those effects, as Melnick and others have shown.

Group Influence on the Forest Service and Bureau of Land Management (Baltimore: Johns Hopkins University Press, 1981), pp. 186–204, 218–29.

41. R. Shep Melnick, *Regulation and the Courts: The Case of the Clean Air Act* (Washington, D.C.: Brookings Institution, 1983), p. 344.

42. Ibid. See also David Schoenbrod, "Goals Statutes or Rules Statutes: The Case of the Clean Air Act," *UCLA Law Review*, vol. 30 (1983), p. 740 (courts' unsuccessful attempt to compel imposition of transportation control plans).

43. See Rabkin, *Judicial Compulsions*, p. 34. The injunction against programmatic litigation in National Wildlife Federation reflects a more realistic assessment of the judiciary's role and can be expected to curb counterproductive judicial ventures.

The same is true of regulatory obsolescence, that is, government's tendency to maintain or expand regulatory programs even when their underlying premises have proven untenable, obsolete or, as the case may be, a demonstrable failure. This proclivity, too, has multiple causes; among them are the tendency of regulatory programs to create intensely interested beneficiaries, bureaucratic inertia, and the institutional design and operating habits of Congress, which favor incrementalism over fundamental reexaminations of existing programs. But, again, the legal presumptions of the environmental era and in particular the preoccupation with capture have played a large role. One-dimensional statutes—reinforced through agency-forcing provisions, legislative "hammer" provisions that impose absurdly tough standards unless the agency regulates within a certain time frame, and similar instruments—were ostensibly designed to keep agencies on their toes. In the end, though, an unrelenting commitment to absolutist values is only another name for regulatory obsolescence: so long as the agency falls short of the stated objectives—and it *always* falls short—there is reason to pursue even clearly counterproductive policies with increased vigor.

Hard-look review reinforced this tendency: its explicit purpose was to freeze agencies into existing policies, for fear that a change of heart might be due to politics and might thwart the will of Congress.[44] *Chevron*-esque deference, as noted, represents a 180-degree departure from these presumptions: it recognizes that regulatory agencies can, should, and occasionally do learn from experience and that transcendental values are no substitute for manageable programs and acceptable results. Whereas hard-look review tended to view the election of a new president, with a new administration and different political priorities, as an unwarranted intrusion of politics into sacrosanct value commitments, *Chevron* views such events as a legitimate source of policy changes.[45] Substantive review rests on somewhat different normative assumptions but also leaves agencies considerable leeway to revise, amend, or abandon obsolete policies.

What is true of the effects of judicial review on policy fragmentation is also true of its effects on regulatory obsolescence: the who of

---

44. See State Farm Mutual Automobile Insurance Co. v. United States Department of Transportation, 680 F.2d 206, 220 (D.C. Cir. 1982) ("sharp changes of agency course constitute 'danger signals' to which a reviewing court must be alert"; citing Joseph v. FCC, 404 F.2d 207, 212 (D.C. Cir. 1968)); ("rescission of [a policy] must be subject to 'thorough probing, in-depth review' lest the congressional will be ignored").

45. Chevron, U.S.A., v. NRDC, 467 U.S. 863, 865–66 (1984) (Stevens, J.).

review may be more important than the how. It has become fashionable to decry the ossifying effects of judicial review.[46] To some extent, these effects occur regardless of whether review is initiated by the beneficiaries or by the objects of litigation. (The tendency of administrators to compile voluminous records, often with an eye not to substantive relevance and accuracy but toward surviving the invariable legal challenge, is an example.) But the principal legal device that produces ossification and obsolescence-inducing fidelity to statutory objectives is not judicial review per se. That device is the citizen suit, which is specifically intended to immunize disembodied values from politics and to ensure that "purposes heralded in Congress are not lost or misdirected in the vast hallways of the bureaucracy," as Judge Skelly Wright put it in an early, precedent-setting case.[47] Usually, the best litigation result that the objects of regulation can hope to achieve is to send the agency back to the drawing board. These delays may be prolonged; in rare cases, an agency may decide to abandon a particular rule-making proceeding. But this outcome is a far cry from the demobilizing effects of lawsuits brought by beneficiaries who demand fidelity to legislative purposes or—what amounts to the same thing—a "right" in the regulatory status quo. Here is the essence of obsolescence: regulatory beneficiaries come to view ostensibly ironclad statutory commitments as entitlements and will not permit agencies to retreat or, for that matter, simply to change their minds.[48]

Examples of beneficiary lawsuits that have frozen agencies into obsolete policies are legion. The decade-long *Chevron* litigation was directed against the EPA's modest effort to introduce flexibility and economic incentives into the regulation of stationary sources under the Clean Air Act, a policy initiative regarded by an overwhelming majority of experts as long overdue, efficient and, if anything, far too limited in scope.[49] Similarly, the so-called Delaney clause, which provides that no food or color additive shall be deemed safe if it induces cancer in

---

46. See, for example, Richard J. Pierce, "The Role of the Judiciary in Implementing an Agency Theory of Government," *New York University Law Review*, vol. 64 (1989), pp. 1239, 1263–65; and McGarity, "Some Thoughts on 'Deossifying' the Rulemaking Process."

47. Calvert Cliffs Coordinating Committee v. Atomic Energy Commission, 449 F.2d 1109, 1111 (D.C. Cir. 1971).

48. See Melnick, *Regulation and the Courts*, pp. 360–87.

49. See, for example, Robert W. Hahn and Gordon L. Hester, "Where Did All the Markets Go? An Analysis of EPA's Emissions Trading Program," *Yale Journal on Regulation*, vol. 6 (1989), p. 109, and Richard A. Liroff, *Reforming Air Pollution Regulation: The Toil and Trouble of EPA's Bubble* (Washington, D.C.: Conservation Foundation, 1986).

man or any animal, has widely been described as an example par excellence of regulatory obsolescence.[50] It was enacted in the late 1950s, before it was known that most substances would prove carcinogenic to some animals at some dosage—without, however, posing any danger to human health. Strict interpretations of the clause have resulted in the withdrawal of perfectly safe additives and their replacement with far more dangerous substances. Public-interest lawsuits, however, have forestalled flexible interpretations or *de minimis* exceptions to the Delaney clause.[51]

The argument just sketched is neither new nor original; it was developed much more fully, over a decade ago, in a law review article by then-Judge Scalia.[52] His opinions in *National Wildlife Federation* and in *Defenders of Wildlife* reflect the insight that citizen suits tend to immobilize agencies or, in constitutional categories, to impede energy in the executive.[53] Both decisions go a long way toward curbing this tendency. One may object that Scalia's view betrays a hidden preference for private orderings, and one may insist that "the interests of regulatory beneficiaries deserve no less legal protection than those of the regulated."[54] But the price of extending equal protection to environmental beneficiaries is obsolescence. This, too, is a trade-off that must be faced.

---

50. For example, Cass R. Sunstein, *After the Rights Revolution,* pp. 88–89, 94–95, 198–99.

51. Public Citizen v. Young, 831 F.2d 1108 (D.C. Cir. 1987), cert. denied, 485 U.S. 1006 (1988); Les v. Reilly, 968 F.2d 985 (9th Cir. 1992), cert. denied, 113 S.Ct. 1361 (1993). Technically, these lawsuits were not brought under citizen suit provisions; however, the generalized "consumer standing" on which the plaintiffs relied is to all intents and purposes indistinguishable from citizen standing.

52. See Antonin Scalia, "The Doctrine of Standing as an Essential Element of the Separation of Powers," *Suffolk University Law Review,* vol. 17 (1983), pp. 881, 884.

53. See *Federalist* No. 70 (A. Hamilton) ("Energy in the executive is a leading character in the definition of good government.").

54. Sunstein, "What's Standing after *Lujan*?" pp. 209, 216–22. See also Cass R. Sunstein, "Interest Groups in American Public Law," *Stanford Law Review,* vol. 38 (1985), pp. 29, 75, n. 200.

6
Environmental Ideology and Real-World Politics

Beginning in the late 1960s and for the better part of two decades, ecological presumptions shaped a comprehensive regulatory system in which American society invested well over $150 billion each year. The results have not been encouraging: environmental regulation is plagued by a woeful lack of coordination, tunnel vision, and paradoxes. The courts, however, have come to recognize that environmentalism's constitutive legal doctrines contributed substantially to these systemic inefficiencies, and they have returned to more traditional, harm-based doctrines that promise to make environmental regulation somewhat more efficient and rational. At a minimum, the new doctrines will curtail the most extreme manifestations of the ecological paradigm—complete deprivations of private property, grants of standing to the world at large, programmatic lawsuits aimed at de facto takeovers of agency programs, and regulations that are unreasonable even by the most modest standards. These results are noteworthy and salutary.

To leave it at this, however, would ignore the broader and perhaps more significant effects of environmentalism's legal demise on the terms of the political debate. When the political debate is as intense and ideological as it has been in the environmental area, its tone matters greatly; in recent years, the tone of the environmental debate has changed dramatically and in a direction that parallels environmentalism's demise in the law. This final chapter argues that the ideological premises embodied in the case law have had a major influence on the terms of the political debate—first in establishing and sustaining environmentalist politics and then, over the past few years, in engineering its demise.

What I have called the ecological paradigm was a purposeful, ambitious judicial effort to construct a legal system wherein environmental needs and values transcended politics and legal boundaries: in view of the infinite complexity of the world and the urgency of environmental problems, there could be no boundaries and fences, no harm-based standing hurdles, and no judge-made barriers to the uncompromising pursuit of environmental values. For two decades, this perspective shaped the dominant legal trends in the seemingly disparate areas of takings, standing, and judicial review—and a regulatory system of extraordinary scope and ambition.

In practice, of course, environmental regulation does not measure up to the fantastic aspiration of organizing a complex world under a single collective value. It is technical and arcane and rarely inspirational; cumbersome and bureaucratic, rather than open and participatory. Nonetheless, the characteristic features of environmental regulation correspond directly to the logic of environmental ideology: environmentalism's boundless ambition is written into absolutist statutes, and the faith in an infinitely interconnected world drives a command-and-control system of daunting reach and detail. While we do not literally pursue environmental values regardless of cost and other real-world constraints, the system operates on the *pretense* that we should do so. The pretense, in turn, cannot be sustained without a powerful ideology, and the courts played an instrumental role in the ascent and subsequent dominance of that ideology. At the beginning of the environmental era, the judicial recognition of environmental values as beyond politics and as deserving of special consideration preceded the rise of environmentalism as a potent political force. Subsequently, the judiciary's endorsement of an extravagant paradigm helped to sustain an equally extravagant regulatory enterprise.

The principal theme of this book has been to show that the *demise* of environmentalism amounts to an equally broad judicial reform project. In each case discussed in the preceding chapters, identical results could have been reached with far less theoretical effort, and arguably without a drastic revision of the constitutive doctrines of the environmental era. Thus, the cases resist a purely result-oriented or, for that matter, a purely partisan interpretation; they reflect a fundamental and lasting change in judicial perspective from collective values to private harms as the lodestar of constitutional and administrative law.[1] The

---

1. For an earlier, similar version of this argument, see Robert Glicksman and Christopher H. Schroeder, "The EPA and the Courts: Twenty Years of Law and Politics," *Law and Contemporary Problems*, vol. 54 (1991), p. 249. To repeat an earlier observation, conservative judicial appointments have played a role in the demise of environmental values and the concurrent resurrection of harm-based legal doctrines; indeed, it would be uncharitable to read Antonin Scalia or Stephen Williams out of a story in which they are the real heroes. But if Scalia and Williams had done no more than to reach the "right" outcomes in environmental cases, the Lucas and Defenders of Wildlife cases would be bores, as would International Union and CEI II. Professor Kathleen Sullivan

purpose and the effect of this shift is to confine environmental aspirations to the ordinary demands of interest-group politics; its tendency and promise is to resurrect private, common-law orderings as an organizing principle of American law.

Over the past five years, environmental politics has undergone a dramatic change, whose themes and the direction bear a striking resemblance to the demise of environmental values in the case law. Environmentalism has lost much of its appeal: environmental groups have lost tens of thousands of members.[2] At the same time, the grass-roots property rights movement has grown by leaps and bounds, and think tanks that advocate environmental policies based on property rights and private markets enjoy increased influence, credibility, and funding. The establishment media have begun to subject environmental policies and their underlying assumptions to critical scrutiny.

Perhaps most important, two decades of firm bipartisan support for aggressive environmental protection were abruptly terminated by an environmental backlash (or reform project, depending on one's perspective) that began in the 103rd Congress and accelerated after the Republican takeover of Congress in 1994. A prohibition on so-called unfunded mandates imposed by the federal government on state and local governments has already become law; concerns over the high costs of environmental mandates played a large role in its enactment. Other proposed reform legislation would subject health and safety standards to stringent cost-benefit requirements. A third set of bills would require that property owners affected by environmental regulation be compensated for their losses, including partial takings. (Similar property rights bills have been enacted or proposed in several states.) The purpose of this agenda—dubbed the unholy trinity by the environmental movement—is not primarily to provide regulatory relief for beleaguered interest groups, although such considerations have played a role. Rather, the purpose is to remedy generic, systemic policy flaws and excesses. The most salient reform proposals repudiate the basic assumptions that have heretofore guided environmental regulation.

These changes in the landscape of environmental politics indicate that the judiciary played an anticipatory, agenda-setting role not only

---

has correctly observed that jurisprudential doctrines (of the sort that drive the Lucas decision, for example) are not easily "mapped" onto political preferences. Sullivan, "Foreword: The Justices of Rules and Standards," *Harvard Law Review*, vol. 106 (1992), pp. 96–98.

2. For membership trends of major environmental groups, see Jonathan Adler, *Environmentalism at the Crossroads* (Washington, D.C.: Capital Research Center, 1995).

in environmentalism's rise but also in its demise: the courts came to view environmental laws as one more regulatory excess in need of pruning, and environmentalists as merely one more interest clamoring for recognition, much *before* this perspective gained significant public or political acceptance—in fact, at a time when Congress, with great public support, was still passing ambitious environmental statutes that carried ecological presumptions to extremes. Only after the courts had articulated the need for balance, limits, and reason in environmental politics did these themes begin to dominate the political agenda.

At the same time, the harm-based doctrines that have replaced the ecological paradigm reflect the reality that courts can nudge the political debate only so far; they cannot transcend the existing social consensus or invent and impose a new one. The doctrines point back to a world of private orderings, property, and contracts—that is, to the world of the common law. But the courts have mobilized common-law assumptions only so far as necessary to confine environmental ambitions to the ordinary workings of interest group politics. A broader reconstruction of common-law precepts would entail a frontal assault not only on environmental regulation but on the entire institutional edifice of interest-group pluralism. Such an endeavor would presuppose a libertarian social consensus or at least a realistic prospect of such a consensus. No such consensus is on the horizon, and a common-law revolution will have to await larger, fundamental changes in the political and ideological climate. But the resurrection of harm-based doctrines may well be a step toward a more thoroughgoing reorientation toward private orderings. At a minimum, environmentalism's demise has introduced a sense of realism into the environmental debate; even this is much more progress than one could have hoped for only a decade earlier.

The Logic of Environmentalism

For some two decades, the ecological paradigm was a powerful influence in American law. Its basic premises—the belief in the transcendence of environmental values and the idea that the law must trump the complexity of the world through a comprehensive regulatory scheme—were not subject to serious political challenge. But these ideas are absurd—so much so that environmentalists themselves do not pursue the ecological paradigm to its natural conclusions. The fact that an ambitious regulatory system built on ecological premises nonetheless flourished for two decades illustrates the power of ideology over law and politics.

Like any ideology, the ecological paradigm arises from—and

routinely recurs to—observations that have a certain undeniable plausibility. There *are* serious and dauntingly complex environmental problems, some of which seem to defy easy solutions within the framework of interest-group politics and the resolution of conventional, individualistic legal disputes. Environmentalism, however, inflates this realistic awareness of difficult cases into an all-encompassing paradigm of complexity. From the existence of vexing—though ultimately tractable—environmental problems, environmentalism jumps to the assertion that liberal institutions and particularly the common law are constitutionally incapable of dealing with complexity and from there to the conclusion that complexity requires a legal system wherein environmental values erase all boundaries and trump all competing concerns and considerations. To put it gently, these inferences are not remotely so plausible as the substantive concerns from which they arise.

The ecological argument of complexity suffers from its partial and selective nature. Our social and moral environment is no less complex than the natural world. Although there is no *theoretical* reason why we should sacrifice all on the altar of complexity in one case while maintaining discipline in the other, we do not seriously aspire to organizing social relations in accordance with doctrines analogous to those that flow from the ecological paradigm. We have refrained from doing so for the excellent reason that such an agenda would do to our personal freedom what the South Carolina Beachfront Management Act did to Mr. Lucas's property, that is, to take it.

Even in its own domain, environmentalism resists the inescapable implications of its own paradigm. Paul Ehrlich, for example, has been arguing for well over two decades that the global ecosystem can support only a fraction of the current world population. (By Ehrlich's lights, China alone has exceeded its carrying capacity by some 500 million people.)[3] Neither these overwrought claims nor their unpalatable implications have prevented Ehrlich from becoming one of the environmental movement's most respected and influential figures; his widely publicized works enjoy a near canonical status,[4] and environmentalists uniformly agree that global overpopulation is a singularly pressing environmental problem. Needless to say, though, environmentalists do not advocate the extraordinary measures that would be

---

3. Paul R. Ehrlich and Anne H. Ehrlich, *The Population Explosion* (New York: Simon & Schuster, 1990), p. 206.

4. One of Ehrlich's earlier books on overpopulation, *The Population Bomb* (New York: Ballantine, 1968), published and widely distributed by the Sierra Club, is one of the most influential contributions to modern environmentalism.

required to bring the world population into an Ehrlichian ecological balance; neither, in the end, does Ehrlich himself.[5] At some point, then, the alleged imperatives of global complexities bump up against compelling liberal commitments that no one, environmentalists not excluded, wishes to jettison.

In this light, the environmentalist indictment of interest-group politics and individualistic legal doctrines loses much of its force. It is quite true that private rights and interests impede the unconstrained collective pursuit of environmental objectives. But such constraints are maintained because a liberal democracy, no matter how rich, operates under conditions of economic scarcity; it cannot and will not subordinate all other social values and interests to environmental needs. Ultimately, environmentalists will not have it any other way, and their insistence on the transcendence of ecological values is a posture—a way of leveraging *their* values over everyone else's values.

The amazing fact is not that this attempt should be made but that it should be so successful. Liberal democratic scruples to one side, the attempt itself to eliminate environmental risk at all costs would be self-defeating. A collective decision to devote the entire GDP to the prevention of accidental deaths, for example, would imply an expenditure of

---

5. Zero Population Growth, for example, an environmental group that propagates Ehrlich's views, is preoccupied with issues of voluntary birth control and especially with access to abortion. In addition, ZPG advocates such policies as recycling and local antigrowth ordinances—measures that seem almost pathetically inadequate in light of the allegedly imminent, catastrophic effects of overpopulation. Ehrlich himself is somewhat more ambitious, and a lot more sinister. While he laments coerced abortions and sterilizations as a "sad thing" about China's one-child policy, he attributes such practices to the country's delay in adopting the program and to peasant resistance. Ehrlich provides an otherwise glowing review of the policy. Among other things, he congratulates the Chinese government on "providing equal rights and education for women" and on its "considerable openness . . . about the successes and failures of the [population control] program, including the human rights abuses that have at times occurred." He concludes from the Chinese experience that "even a very strong program of population control, pushed by a repressive government on a regimented society, can fail if a nation starts too late." Ehrlich, *The Population Explosion*, pp. 205–8. But Ehrlich's entire book is dedicated to the proposition that it is already too late, or nearly so—which raises the question of what a *successful* population control policy would require. In the end, even Ehrlich distances himself from coercion and expresses the "hope" that the world will heed his warnings before governments conclude that only totalitarian measures will suffice to avert global annihilation. Ibid.

some $55 million per life saved[6]—substantially less than the price of regulations that are routinely advocated and defended by environmental interest groups (for example, in *Corrosion Proof Fittings*). No money would be left for other causes that are near and dear to the environmentalist heart and purportedly central to the preservation of a complex and fragile world—the preservation of endangered species, for example, each one of which is also presumed to have an unlimited claim on the national wealth.

Then, too, life is much more complex than is suggested by the thought experiment of devoting the entire GDP to environmental objectives. People eat and drink, work and play, build houses and raise children. Many of these activities consume wealth that would otherwise be available for environmental objectives; all of them entail a danger of death, often far in excess of levels environmentalists consider acceptable. But we insist on pursuing our ordinary activities, and we must do so unless we were to drop dead.[7] Conversely, many of our ordinary activities produce the wealth that enables us to pursue environmental objectives in the first place; thus, rather than trumping everything, environmental values must at the least be balanced with the need to maintain a productive economy.[8]

Environmentalism responds to these complexities in the only way possible: it carves the world into discrete issues and pursues them in abstraction from their larger context—one pesticide, one emission source, one owl at a time. Environmentalism insists on the uncompromising pursuit of these fragmentary objectives even when the broader consequences to health, safety, and environmental quality are demonstrably adverse: so long as more stringent CAFE standards reduce car emissions, their tendency to kill people does not matter. Having exalted a wholly ideological notion of complexity, environmentalism responds to real-world complexities by ignoring or denying them.

---

6. W. Kip Viscusi, *Fatal Trade-Offs: Public and Private Responsibilities for Risk* (New York: Oxford University Press, 1992), p. 5 (1990 figures).

7. The clear import of Paul Ehrlich's theories is that the suicidal are doing us a favor by acting on their impulses. For reasons mentioned, however, environmentalists resist this conclusion.

8. Actually, the fact that prosperity is a prerequisite for the pursuit of environmental objectives drives toward the conclusion that rising levels of wealth are not only consistent with but in fact the surest route to a better, healthier environment. The most compelling and accessible presentation of this "wealthier is healthier" argument is Aaron Wildavsky, *Searching for Safety* (New Brunswick, N.J.: Transaction Books, 1988).

The point here is not to convict environmentalism once and for all on the charges of extravagance and incoherence or to declare it incapable of a thoughtful, circumspect defense against these charges. Rather, the point is that the central tenets of the ecological faith systematically push toward absurd and contradictory conclusions and that *these tendencies have in no way diminished environmentalism's political force*. We do not pursue the ecological paradigm to its logical conclusion, any more than environmental ideology pursues itself to its own conclusion. We do not literally spend the entire GDP on a single endangered species, even when the Supreme Court tells us that we must do so. But we act and regulate *as if* we were committed to doing so—for example, when we ban pesticides at a cost of $100 million per life saved. We do not literally regulate everything and at once; yet such statutes as the 1990 Clean Air Act Amendments—a 600-page-plus compendium of standards, deadlines, and requirements pertaining to the composition of fuels, the sale of cars, the emissions from dry cleaners, lawn mowers, and barbecue lighters, and technological specifications for each valve and pipe in each factory in the United States, among countless other matters—are comprehensible only on the theory that we must do so. We do not regulate literally every private action as a potential assault on the ecosystem, but car drivers, smokers, landowners, and (elsewhere on the globe) the objects of China's one-child policy can attest how far we have gone in subjecting heretofore personal conduct to regulatory controls that are predicated on the ecological paradigm. We would not countenance or even consider such extraordinary impositions were it not for the force of ideology.

It may seem strange that an ideology that tends toward indefensible results and internal contradictions should have such influence on politics and policy. Environmentalism, however, shares this gravitational pull with all ideologies. The point is most obvious in legal and political systems that are dominated entirely by a single ideology: although Communist countries, for example, have never lived up to their official dogma, Marxist ideology has been far from irrelevant to their operation (which is why it was never officially discarded). Contemporary, less comprehensive (and oppressive) ideologies follow the same pattern: civil rights advocates refrain from imposing proportional hiring quotas on the Boston Symphony Orchestra; yet the disparate impact theories that would compel such a result dominate employment law. Feminists adamantly deny any intention of forcing unisex bathrooms on the country—all the while insisting that institutions based on distinctions of sex (such as all-male colleges) are just as odious as those based on racial distinctions.

Ideologies, then, always point far beyond what their advocates are prepared to defend. Environmentalism is no exception to this pattern; it is, in the end, an extravagant pretense. But environmentalism is not a *mere* pretense: for two decades, it sustained a regulatory regime that could not be sustained on any other basis.

The Logic of the Common Law

When the law follows ecological dictates, it erases all limits to the collective pursuit of values—property boundaries, standing barriers, and the requirement of reasonable results. The doctrines that signal the demise of the ecological paradigm, in contrast, reestablish these boundaries. In each of the three areas examined in this book—takings, standing, and judicial review—the boundaries are marked by the notion of tangible, manifest, particularized harms that belong to somebody. A genuine harm to a segment of the outside world must exist before private expectations to the enjoyment of property may be limited. Real harms to the plaintiff are a requisite for standing to sue. Tangible costs and benefits—as distinct from absolutist aspirations—shape the exercise of judicial review. This notion of harms reintroduces a limit to the governmental pursuit of environmental values and to the authority of private parties to compel the exercise of state power against other private parties. In so reestablishing private spheres of action beyond the reach of environmentalist values, the concept of harms reverts to the logic and the basic premises of the common law.

By *common law* I do not mean the *historical* common law as it existed at the time of Blackstone or at the end of the nineteenth century. Rather, I have in mind the basic logic of a legal system whose principal purpose lies in protecting private orderings. Such a system guarantees robust individual rights to exclude others (property); provides avenues for voluntary exchanges (contracts); and protects against aggression by outsiders (torts). Its fundamental maxim is "keep off." As a historical matter, the common law always made exceptions to these general principles; in the heyday of judicial respect for private orderings, for example, the Supreme Court permitted states to prohibit the production and sale of alcoholic beverages, even though the prohibited transactions occur among consenting adults and are not harmful per se.[9] But the historical common law and the vision of private orderings are sufficiently congruent to render the terminology not altogether confusing.

Because of this congruence with private orderings, statist ideolo-

---

9. See, for example, Mugler v. Kansas, 123 U.S. 623 (1887) (denying compensation for state's confiscation of private brewery under a statute prohibiting intoxicating beverages).

gies have always viewed the common law as a particularly stubborn and odious obstacle to collective aspirations. And, uniformly, ideological attacks on the common law have revolved around the proposition that the world has become too complex for private management. A century ago, corporate capitalism was thought to render common-law doctrines obsolete. The New Deal assumed that a modern mass society and new technologies compelled a drastic revision of constitutional and common-law arrangements. Only a few decades ago, pervasive racism was presented, successfully, as a compelling argument against freedom of contract. In our own day, feminists clamor for a drastic revision of common-law and constitutional (and, indeed, criminal law) protections as the only means of overcoming structural obstacles to sexual equality. The connecting theme is collectivism: time and again, collective political schemes have been proffered as the only functional response to new and unique problems the common law had either suppressed or simply not thought of. "Complexity" has been statism's perennial battle cry. Environmentalism sounds it yet again, albeit in a particularly shrill tone.

The ideological argument typically proceeds from a picture of the common law as a rather primitive legal system that is based on a vision of autonomous, atomistic individuals who relate to each other through simple, one-on-one transactions and an occasional fist in the face. Anything much more complex allegedly exceeds the common lawyer's grasp and comprehension. From this starting point, the argument naturally leads to the conclusion that only political orderings can account for the complexities of modern life. Even in its more sophisticated forms, though, this perspective rests on a misunderstanding of the common law—or, less charitably but perhaps more accurately, on an ideologically inspired caricature. Like any sensible theory, the common law *starts* with the simple, general cases. The choice of the starting point implies a presumption in favor of private orderings (though no more so than the decision to put complexity front and center implies a theoretically implausible and hence even more arbitrary preference for statist arrangements). But nothing in the basic logic of common-law arrangements precludes their extension to more complex cases.

In fact, so far from ignoring complexity, the common law presupposed it—and offered private orderings as a solution. The contrast between the common law and the ecological perspective lies not in the recognition of complexity but in the different responses to that recognition. The basic intuition of the common law is that precisely *because* the world is complex, it needs simple rules that allow it to be managed not through collective, centralized, one-size-fits-all arrangements but in small chunks and by individuals who are likely to get the results right.[10] Of course, private actors will make mistakes. But so will the government, and the costs of *its* mistaken decisions will be much higher. Private actors learn, as long as—and because—the rules ensure that they bear the consequences of their mistakes. The government, having no such incentive to learn and no competitors who could put it out of business, tends to lurch from one inefficient scheme to the next.

With respect to problems we now think of as environmental, the common law was, in fact, quite attuned to complexities—externalities, aggregate effects, multiple causation, unquantifiable risks, and indirect but nonetheless real effects. Although the common law may not have thought of the planet as Spaceship Earth, it did deal, as it had to, with complex social systems and ubiquitous and subtle externalities. Private orderings are not suitable to every conceivable problem. In some settings, the transaction costs entailed by private arrangements may be sufficiently high to render collective, political solutions more efficient. Market failures and collective action problems may require government action; this is why the government may "take" private property for public use, provided it compensates the losers. But there is no reason why the marginal and unique cases should overwhelm the larger vision.

Private orderings depend on boundaries and fences; nothing can be sorted, and nothing can be arranged, if the world is one vast common pool. And, so, the common law sorted things into mine and thine, protected private property and exchanges, and provided protection against force, fraud, and aggression. These operating principles, which are the basis of a free society, would crumble if everything were to become everyone else's business. In *this* sense, the common law suppressed complexities: it cut off chains of causation, distinguished legal injuries from unredressable harms, and separated torts and invasions from the ordinary risks and the background noise of life.

As a historical matter, these distinctions had a certain arbitrariness. But this criticism—reflected, for example, in the complaints about the idiocy of common-law life in Justice Harry Blackmun's *Lucas* dissent—misses the point: some arbitrariness at the margins matters little so long as the underlying principle remains clear. It may in some sense seem arbitrary, for example, to provide a legal remedy for injuries from trespass but not for potentially more substantial injuries from economic competition. But the appearance of arbitrariness dissolves from the vantage point of the underlying principle: private property requires protections against invasion; private markets could not work

---

10. An ingenious modern version of this argument is Richard Epstein's *Simple Rules for a Complex World* (Cambridge: Harvard University Press, 1995).

if successful competitors had to compensate the losers. Similarly, the common law understood employment relations as private, voluntary agreements that had to be preserved regardless of the harms they might entail in certain cases. Individuals are free to contract into arrangements that may entail high risks; they will usually demand and obtain correspondingly higher compensation. (The substantial salaries paid to football players, boxers, and firefighters in the Persian Gulf illustrate the general principle.)[11] The flip side is that there can be no common-law remedy for risks that are voluntarily assumed under the terms of an employment contract. The corollary constitutional proposition is the logic of *Lochner*:[12] absent a genuinely compelling reason, government may not regulate so as to preempt or abrogate the private right to contract for employment, including risky employment. Again, the preference for private orderings flows from the recognition that the world is a messy place: since risks are uncertain, and private risk preferences vary greatly, risks are generally best managed by those who are directly affected by them.

American law has moved a long distance from the vision of common law. Some six decades ago, exclusive, individual rights gave way to interest-group entitlements, and private orderings were replaced with interest-group politics as the basic paradigm of the law. This shift, in turn, paved the way for environmentalism's ascent. The transition from rights to entitlements to transcendental values, however, has now been reversed: the harm-based doctrines that spell environmentalism's demise in law point back toward the logic of private orderings. Much like the fantasy of the ecological paradigm, the realism of the common law has a certain gravitational force—even if the courts, for the time being, are reluctant to follow its attraction.

The Ambivalence of Interest-Group Politics

Since the New Deal, American law has reflected the basic presumptions of interest-group pluralism. These presumptions are reflected, for example, in the interpretation of the takings clause, the contracts clause, and the equal protection clause; the interpretation of the structural provisions of the Constitution, such as the powers of Congress under the commerce clause; the Administrative Procedures Act; and the constitutional and statutory construction of standing to sue. The

---

11. For a concise account of this compensating wage differential, see W. Kip Viscusi, *Risk by Choice* (Cambridge: Harvard University Press, 1983), chap. 3.
12. Lochner v. New York, 198 U.S. 45 (1905) (striking down state law regulating maximum working hours in the bakery industry).

legal system is built not around rights but around interest-group claims; the common law's maxim ("keep off") is replaced with something along the lines of "come on in; tell us who you are and what happened to you; and if it's serious, we'll see what we can do." The operating principles are not markets, property, and contracts but collective bargaining, legislative logrolling, and bureaucratic management.

The cartel arrangements of the New Deal, briefly discussed in chapter 1, are the classic example of a system that replaces the freedom to compete with a limited freedom from competition, and markets with politically managed regimes. Other, more recent examples come readily to mind. Civil rights protections against discrimination, for instance, may look like individual rights. But the right to nondiscrimination exists by virtue of an employee's membership in a legislatively defined class (race, sex, and so forth). Similarly, uniform workplace safety standards—imposed by OSHA or determined by collective bargaining—replace freedom of contract with a right to a safe workplace; again, the reference point is workers as a class, not the individual worker who might well wish to contract for riskier employment, provided the price is right. And, to pick an example outside the labor context, the property owner's common-law right to pursue his claims in court (and the corollary right *not* to be a party to a lawsuit without his explicit consent) is replaced with a right to legal representation by a collective association, provided the group is broadly representative of the interests for which it claims to stand.[13]

Such interest-group entitlements cannot simply be superimposed on a common-law regime. They are often parasitic on the common law; the right to nondiscrimination in employment, for example, is often viewed as an expansion of freedom of contract. In fact, however, the new entitlement replaces the bilateral, reciprocal freedom of contract with the employee's unilateral right to demand fair treatment, while constraining the employer's more fundamental right *not* to deal. In other words, each new entitlement entails a corresponding contraction of the individual's right to exclusive control over his labor or other property.[14] The scope of this expansion and the corresponding contrac-

---

13. Hunt v. Washington State Apple Advertising Commission, 432 U.S. 333 (1977).

14. The "labor" point is important: as just noted, uniform OSHA standards do not simply cost the employer; they also prevent individual workers from accepting riskier jobs at higher pay. The question is not whether uniform standards are "better" or more efficient than a regime of individual, reciprocal rights (although the evidence tends to indicate that they rarely are). The point is that collective standards and entitlements cannot be superimposed on a common-law regime without a cost to the latter.

tion are left, as they must be, to the collective pursuit of political interests. The legal game moves from reciprocity to redistribution.

Once the theoretical baseline has shifted from private to political orderings, though, the ground is prepared for a slide into disembodied values and their unconstrained collective pursuit: it becomes difficult to explain why politics should be confined to redistribution games among competing *interests*, instead of facilitating the common pursuit of nobler aspirations. Especially when interest-group politics begins to look politically unappealing (as it did in the 1960s), it seems compelling to orient the law toward genuinely common purposes instead of partial interests. The rise of the ecological paradigm reflects this dynamic. If Congress may replace private property rights with political regimes (the better to vindicate the public interest), what gives the courts the right to second-guess Congress as to how far it may go? If Congress can recognize harms that affect interest groups, why can it not recognize harms that affect the public at large? If interest groups capture administrative agencies, why should the courts not counteract this tendency and reassert the public purposes Congress had in mind? The seeming lack of ready answers to such questions facilitated the triumph of environmental values.

And yet the trajectory of the law suggests that the transition from harms to values cannot be accomplished quite so easily. After the New Deal had compromised common-law orderings, it took three decades before interests and entitlements drifted off into values. And even in the heyday of values-oriented jurisprudence, the difference between entitlements and values was never quite erased. The courts recognized new entitlements to food stamps, welfare benefits, and due process and new rights to privacy and proportional representation in schools and employment. The push for a constitutional right to a healthy environment—no less plausible at first sight than other newly minted rights—failed spectacularly, with several courts declaring that there was no such right.[15] Conversely, cases in which the courts discovered new entitlements were suffused with rhetoric about public values;[16] yet the courts never declared, as they did in the environmental area, that "any citizen" had a right to enforce those values "whatever the cost." The ecological paradigm of values-oriented jurisprudence never became a paradigm for the law at large.

---

15. An instructive discussion is Gerald N. Rosenberg, *The Hollow Hope: Can Courts Bring about Social Change?* (Chicago: University of Chicago Press, 1991), p. 273.

16. The best treatment of the subject is R. Shep Melnick, *Between the Lines: Interpreting Welfare Rights* (Washington, D.C.: Brookings Institution, 1994).

The reason for this discrepancy is not that the courts cared more about one set of entitlements or values than the other. Rather, there is an ineradicable difference between public values and private harms, even if the line is blurry at times. Harms and entitlements always belong to somebody and separate that somebody from the world at large. No entitlement is boundless; we provide the poor with food stamps but not with first-class dinners in a lavish environment. And every entitlement suggests a corresponding obligation on somebody else's part. Entitlements, in short, invariably define limits and segment the world. This is why transcendent ecological values cannot be conceptualized in terms of entitlements: the attempt to do so would entail the absurd conclusion that everyone is entitled to everything against everyone else.[17]

The harm-based doctrines that spell the demise of the ecological paradigm reflect the insight that universal claims of the sort just mentioned are no less absurd simply because they are couched in the language of values. In a sentence: it cannot be right that everyone has standing to make unlimited claims on everyone else's private possessions. There must be barriers—in the form of injury-in-fact, of reasonableness, and of property rights—that determine who can claim how much against whom.

Thus, in erecting barriers to the imposition of values, the courts' harm-based doctrines revert to the logic of entitlements and interest-group politics. But the doctrinal shift, once underway, is not easily arrested. Interest-group pluralism implies a primacy of political over private orderings and, for this reason, pushes toward values. But interest-group pluralism *also* insists on barriers and boundaries, and once this realization sets in, the logic of interest-groups gravitates back toward the common-law world of private orderings. The notion of tangible harms as a limit to state power points back to individual, exclusive rights.

Shorn of these near-Hegelian abstractions, the point emerges quite clearly in Cass Sunstein's discussion of the "concrete, de facto" injury-in-fact requirement that forms the baseline of the standing analysis in *Defenders of Wildlife*. Any legal theory that revolves around harms, Sunstein points out, presupposes some principle that distinguishes

---

17. If you doubt it, consider again the gnatcatcher's entitlements under the Endangered Species Act: the act confers on every imperiled species the functional equivalent of an absolute entitlement to all the habitat it might need (and without having to pay compensation). See chapter 1, p. 12 n. 24. The functional equivalent is no less absurd than the real thing, but it does not look equally absurd, at least not at first sight.

trivialities from the harms that count for legal purposes. *Defenders of Wildlife* purports to find this principle in the concreteness or directness of the harms (hence, injury *in fact*). Sunstein shows persuasively, however, that harms per se cannot tell us whether they should or should not count; the distinction requires some *prior* understanding of how the world works or ought to work.[18] One possible source of such an understanding is the legislature: one could say that harms should count when and because the Congress has said they should count. According to this theory, though, the concept of harm loses its independent force and might as well be abandoned; we are back to the world of legislative supremacy and unconstrained values.

Thus, if the injury-in-fact requirement is to serve its stated purpose of foreclosing the flight into legislative values, it must be based on a nonlegislative source, and this can only be the common-law understanding of how markets and *private* orderings work.[19] And, lo, as Sunstein observes, the standing analysis gravitates back toward the sort of harms that are analogous to or familiar from the common law. Perhaps the best example is the principle—prominent in *Defenders of Wildlife*—that the objects of regulation (that is, the regulated industries) have standing simply by showing that they are affected by a challenged rule or decision, whereas "much more is required" of the beneficiaries of regulation, such as environmental groups: the holding makes sense only against the background of the common-law understanding of rights as a sphere of private autonomy.[20]

To be sure, harms that are analogous to common-law claims are not *identical* to exclusive rights.[21] In reverting to harm-based doctrines, the courts have fallen far short of a full-fledged reconstruction of common-law precepts. For the time being, they have mobilized common-law–like conceptions only so far as is necessary to confine environmental aspirations to interest-group politics: interests may still roam, even

---

18. Cass R. Sunstein, "What's Standing after *Lujan?*" Of Citizen Suits, 'Injuries,' and Article III," *Michigan Law Review*, vol. 91 (1992), pp. 188–91, 226. Recall from the preceding discussion of the common law that nothing in the harms themselves explains why there is a legal remedy for trespasses but not for—often more severe—injuries from competition.

19. More precisely, it must be based on a theory that views private orderings as the baseline of constitutional interpretation. The difference is of considerable interest, but it does not affect the present analysis.

20. Sunstein, "What's Standing after *Lujan?*" p. 190. See also Sunstein, "Standing and the Privatization of Public Law," *Columbia Law Review*, vol. 88 (1988), pp. 1432, 1435–36; and p. 56 above.

21. As noted in chapter 3, pp. 56–57, Sunstein tends to slight this difference.

though values may no longer do so. Except in the narrow context of total wipeouts in land use cases (where the tangible harm is so great as to require categorical rules), *Lucas* leaves property rights subject to political bargaining and judicial balancing. Under *Defenders of Wildlife*, environmental groups can no longer act as the faithful guardians of environmental values ordained by Congress. But they are *interests* like any other interest, and so long as they look and act like any other special pleader, *Defenders of Wildlife* entitles them to equal concern, respect, and standing. Aspirational statutes no longer deserve special judicial solicitude, but, especially in their deferential *Chevron* mode, the courts will still pay great respect to successful interest-group bargains.[22]

It remains true, nonetheless, that harm-based doctrines point further toward a reconstruction of common-law premises than the courts are at present prepared to go. *Lucas* is almost entirely parasitic on the common law; in fact, the aggressive reassertion of common-law doctrines in the context of complete wipeouts compels the question why those doctrines should be inapplicable outside that context, and *Lucas* provides no good answer. The drift toward private orderings is somewhat less obvious in the judicial review cases. But from the requirement that governmental risk-regulation must be reasonable (in the sense of doing no more harm than good), it is not a huge step to resurrecting the maxim that statutes in derogation of the common law must be narrowly construed—that is, to a judicial review that uses private orderings as the baseline of reasonable legislation and regulation.[23]

---

22. Even the more aggressive substantive review cases fall well short of suggesting that the government must have a compelling reason before it regulates and, in so doing, preempts the individual risk or cost-benefit assessments of private citizens. See, for example, CEI II, 956 F.2d at 327 ("*When* the government regulates in a way that prices many of its citizens out of access to large-car safety, it owes them reasonable candor.") (emphasis added). The requirement that agencies *explain* their actions is far from questioning the government's authority to act.

23. The suggestion is far from absurd. The aforementioned "compensating wage differential" for risky jobs, for example, not only allows workers to sort themselves into more or less dangerous jobs in accordance with their individual risk preferences but also provides employers with a powerful market incentive to reduce workplace risks to the socially optimal level. On these grounds, one could argue for a presumption that regulation of the remaining risks must be based not on unsubstantiated congressional guesses or easily manipulated cost-benefit assessments but on a persuasive showing of a manifest market failure. See Viscusi, *Risk by Choice*, chap. 3, esp. p. 58. Viscusi concedes that market failures—for example, the fact that employees may not be informed of the true risks they are assuming—might justify government inter-

The drift toward common-law presumptions stems from the recognition that the world cannot be managed as a seamless web; legal regimes must reduce rather than mimic complexity. Having relearned this lesson, though, why should we draw the line at the point where complexity is managed by interest groups and bureaucrats—with all the inefficiencies and shenanigans that such political arrangements ordinarily entail? Put differently, and perhaps more to the point: the impulse to rein in values reveals a distrust of an unconstrained political process. And once this distrust sets in, it becomes difficult to dispel the suspicion that not simply the values but also the politics should be constrained. Even when confined to the pursuit of interests, our political commons produce tragedies; our private backyards, as a general rule, do not. The demise of environmentalism has reopened this long-forgotten challenge.

The Effects of Ideology

Legal paradigm shifts do not occur in a vacuum. The rise of the ecological paradigm occurred at the tail end of the long postwar era of rapidly rising prosperity, which heightened the demand for environmental amenities; it also reflected an increasing social distrust of experts, elites, and big institutions and a corresponding yearning for more participatory, democratic politics. Similarly, the demise of the ecological paradigm unfolded against a background of a growing anti-regulatory mood among the public and a renewed interest in private orderings among economists and legal scholars. Especially in retrospect, environmentalism's rise and demise in the law may look like mere reflections of these social and intellectual trends.

Broad social trends, however, almost always explain too much and too little at once. On the one hand, they rarely entail all the consequences they could or should entail; on the other hand, they tend to attach little significance to the mechanisms through which abstract ideologies and general public moods are transported into law, institutions, and policies. However marginal such mechanisms may appear from a broad-sweep-of-history perspective, though, ideologies cannot become dominant without them. Both in developing the ecological paradigm and in engineering its demise, the courts provided just such a mechanism or channel through which ideological presumptions assumed political dominance.

The judiciary's ability to shape politics and its institutional context

---

vention, though probably of an information-providing rather than a standard-setting kind. Ibid., chap. 5, esp. pp. 84–87.

in accordance with ideological precepts is not a matter of raw political power. Except under highly unusual circumstances—such as a public consensus on issues of central importance—courts cannot bring about fundamental social or institutional reforms.[24] Courts are not designed for this purpose; they cannot force Congress to pass particular laws or force voters to change their minds about public policy. Because of their institutional design and incentives, however, courts do have the ability and, at times, the inclination to reshape the terms of the political debate by playing an anticipatory, agenda-setting role.

Judges are generalists, not experts. They are expected to explain the reasons for their decision in the form of a principled, coherent argument and to consider how any given set of rules or presumptions will play out in the next case ahead, which may involve a different cast of actors and different policy considerations. Thus, courts are by institutional design somewhat more doctrinaire than the political branches, whose deliberations tend to be oriented more toward expediency and acceptable results. Judges are *less* likely than bureaucrats to get absorbed in policy details and *more* likely to proceed from a series of observations to a general theory or paradigm—from repeated bureaucratic foot-dragging to capture theory, for example. Shifts in the intellectual edifice are therefore discerned earlier and more easily in appellate decisions than amid the noise and clatter of the political process. By the same token, courts are more likely—again, compared with politicians and bureaucrats—to rethink the basic premises of regulation when things go wrong. Especially when policies appear to fail repeatedly, in similar ways, and for similar reasons, courts will be inclined to rethink the basic assumptions when politicians, unable to as-

---

24. See especially the careful study by Rosenberg, *The Hollow Hope*. The widespread impression of the Supreme Court as an engine of social and institutional reforms—dubbed the "dynamic court view" by Rosenberg—rests largely on the Court's activist jurisprudence on hotly contested issues, notably, race relations, crime, abortion, and the separation of church and state. Obviously, judicial rulings on these issues have greatly affected public policy. As Rosenberg shows quite persuasively, however, there are good reasons to doubt that the Supreme Court has been able to impose its will—independent of a public consensus—even on these issues. Largely consistent with this view, Alexander M. Bickel, *The Morality of Consent* (New Haven: Yale University Press, 1975), has interpreted the Supreme Court's jurisprudence as a way of gaining democratic legitimacy by anticipating a social consensus on central political issues. As Bickel observed, the Court guessed right on desegregation. A wrong guess on crime and probably one on abortion explain why the Court eventually retreated from its aggressive positions against the death penalty and in favor of virtually unrestricted abortion.

semble a sufficiently broad coalition for reform, are still looking for quick fixes.

Consistent with these general observations, courts tend to look particularly powerful and inspirational in the early stages of a political movement—especially in the eyes of that movement. To an extraordinary extent, the nascent environmental movement relied on the courts to validate and promote its agenda—not simply for tactical reasons but from a conviction that only the courts could and would rise above partisan interests, "save the whole society," and give "shape and reality and legitimacy" to "the kindest and most generous and worthy ideas."[25] However inflated, these expectations were not entirely wrongheaded. Environmental citizen suits, for a notable example, were a judicial invention. Their creation by the D.C. Circuit transformed the National Environmental Policy Act and its requirement of environmental impact statements for major public works projects—intended by Congress as a nice but inconsequential piece of legislation—into a powerful tool for the environmental movement.[26]

After a few years, agencies learned to prepare impact statements that could pass judicial muster, and environmentalists learned that lawsuits and procedures cannot turn development agencies into ministries for global survival.[27] Measured by the environmentalist expectation of comprehensive institutional reform, NEPA turned out to be not much more than a symbol. But the environmentalist disappointment over this result betrays a lack of realism and, moreover, fails to grasp that even symbols can have a powerful impact on the political debate. With each environmental impact statement, development agencies had to pay tribute to the alleged supremacy of ecological values, thus progressively elevating an absurd presumption to an unquestioned and unassailable principle. At the same time, NEPA litigation tended to elevate environmental litigation groups—which were storefront operations at the time—to the status of equal players in the regulatory process. Although the judicial invention of citizen standing did not literally *create* these groups, it did create a critical opening for them. The politi-

---

25. Rosenberg, *The Hollow Hope*, p. 273 (quoting environmental activists and legal scholars).

26. Paul J. Culhane, "Procedural Change and Substantive Environmentalism," in Theodore J. Lowi and Alan Stone, eds., *Nationalizing Government* (Beverly Hills: Sage, 1978), pp. 204–12.

27. See, for example, Joseph L. Sax, "The (Unhappy) Truth about NEPA," *Oklahoma Law Review*, vol. 26 (1973), p. 239. See also Richard A. Liroff, *A National Policy for the Environment* (Bloomington: Indiana University Press, 1976), still the best account of NEPA's history and impact.

cal process, preoccupied as it is with the tangible concerns of real constituencies, pays little heed to small groups, especially when they explicitly advocate an abstract principle that transcends anybody's actual interests. Environmental citizen suits, in contrast, rest on the premise that the abstract principle is more important than anybody's partial concerns. By granting environmental groups the privilege to assert the general principle, the courts conferred on them legitimacy, which the groups subsequently leveraged into political clout.

As this example illustrates, the principal effect of citizen standing—and of the hard-look doctrine and of absolutist judicial construction of legislative mandates—was not to ensure sound policy results but to translate ideological preconceptions into doctrines and operational principles that, once in place, drive the political process and immunize the regulatory regime against political challenges. For two decades, through economic recessions and expansions, in Republican and Democratic administrations, Congress passed a flurry of environmental legislation—with each successive statute a closer approximation of environmentalist orthodoxy. Before the D.C. Circuit's reinvention of NEPA, no legislator had even heard of citizen suits; on the first occasion thereafter, Congress wrote the first citizen-suit provision into the 1970 Clean Air Act (over no more than token opposition) and more expansive provisions into subsequent enactments. From mandating "best available technologies" and standards regardless of cost, Congress moved to more specific and more demanding standards. From requiring EPA compliance with tight deadlines (which the agency missed), Congress moved to tighter and more numerous deadlines (which the agency also missed). Having passed 80-page statutes, Congress proceeded to pass 600-page amendments. Having prohibited the discharge of pollutants into navigable waters, Congress proceeded to bludgeon agencies into designing, under color of that authority, a wetlands preservation program that wiped out private property titles across the country. And so on.

Ecological preconceptions did not carry the day because they are widely shared or believed. To the contrary: as R. Shep Melnick has shown, for example, legislators take great care to ensure—through budgetary means or oversight hearings—that the EPA does *not* enforce absolutist environmental statutes, for the obvious reason that full enforcement would entail utterly unacceptable consequences.[28] Rather, ideological preconceptions create incentive structures that continue to drive the legislative and regulatory process long after the original in-

---

28. R. Shep Melnick, "Pollution Deadlines and the Coalition for Failure," *The Public Interest*, no. 75 (spring 1984), p. 123.

spiration has been forgotten and the results have proven absurd. No legislator is sufficiently starry-eyed to believe that another round of clean air mandates will actually produce clean air over Los Angeles. But every congressman, every senator welcomes the opportunity to vote "for clean air"—all the more so since the EPA's certain and intended failure to enforce the law provides a subsequent opportunity to rail against bureaucratic "lawlessness" and "sweetheart deals" with corporate polluters. Thus, the combination of ideological presumptions and political incentives immunizes the regime against political challenges. The more demagogic the premise, the greater its political use.

Even when the game breaks down, its ideological premises prove impregnable. So, for example, lawsuits over the Tennessee snail darter—a subspecies of a perfectly ordinary fish—threatened to derail the completion of the Tellico Dam, a project in which Congress had invested over $100 million. Instead of reexamining the premises of a statute that entailed such a bizarre consequence, Congress took the easier path of exempting the Tellico project from otherwise applicable law. In the same fashion, Congress has exempted—or has prompted regulatory agencies to exempt—saccharine (a mild carcinogen) from the Delaney clause, the Trans-Alaska Pipeline from the Wilderness Act, experimental AIDS drugs from the Food and Drug Administration's cumbersome approval process, and banks and municipalities from Superfund's ruinous liability provisions. Environmentalists routinely denounce such exemptions as the triumph of politics over ideology. But the fact that legislators adhere to ecological pretensions even in the face of absurdity and occasional train wrecks indicates that the opposite is more nearly true.

These dynamics explain why the impulse for environmentalism's demise originated in the courts: since the terms of the debate foreclose a realistic possibility of legislative reform, courts are more likely—or, more precisely, less unlikely—than legislators to reexamine the ideological premises of environmental regulation. This reexamination was facilitated by several factors. Environmental policy failures did not look like a series of individual policies gone awry. Rather, many policies and programs appeared to fail in similar ways and for similar reasons; this impression was confirmed by a near consensus among the experts who are supposed to know such things. Moreover, the failures seemed to be closely linked to doctrines and presumptions that the courts themselves had put in place and which they were therefore in a position to reverse.

The signal importance of this reexamination and of the doctrines that have developed in its wake lies not in the tangible effects on public policy, no more than NEPA's importance lay in its immediate impact on public works projects. Rather, in reexamining the premises of the regulatory edifice, the courts have put new questions—or old but long-forgotten questions—on the political agenda. Justice Scalia's *Lucas* opinion provides perhaps the clearest illustration.

Property rights advocates have celebrated *Lucas* as a new birth of freedom. In light of the limited holding and effects of the case, this enthusiasm looks like part delusion, part political spin control (as does the environmental movement's attempt to portray *Lucas* as utterly meaningless).[29] But there is something to the view of *Lucas* as a breakthrough, since the opinion so pointedly rejects the central tenet of the ecological faith. Instead of being taken in by environmental rhetoric about complexities, *Lucas* treats it as a subterfuge for extortion. Once this specter is raised, it poses inconvenient challenges to ecological aspirations. Granted the importance of those aspirations, why must we ruin family farmers and hapless owners of retirement homes in their pursuit? Especially if ecological values are so important and so widely shared, why would the public not be willing to pay for the requisite measures, as the takings clause requires?

Lucas, in short, puts environmentalism on the defensive not because its holding precludes badly needed land controls (it does not) but because it poses questions that environmentalism had successfully suppressed and does not wish to answer. The same is true of the standing cases: in what sense can environmental interest groups be said to represent the universe, and why should they alone be permitted to raise such claims? And so, too, with judicial review: if public health is as important as we all believe it to be, why not make an effort to do more good than harm? Why should we cling to safety standards that are known to kill people?

One can conceive of ecological answers to such questions: oceanfront homes, environmentalists may say, *do* exacerbate the problems arising from global warming, provided the complex causal chains are understood. While asbestos abatement may on balance have killed people, it *would* be beneficial if we invested sufficient funds in state-of-the-art removal operations. The Natural Resources Defense Council *should* have standing to challenge, on behalf of its California members, the government's failure to prepare an environmental impact statement that would address the heretofore ignored possibility that lower CAFE standards would contribute to higher car emissions, which would exacerbate global warming, which would in turn drown the Golden State

---

29. See Richard Lazarus, "Putting the Correct Spin on *Lucas*," *Stanford Law Review*, vol. 45 (1993), p. 1411.

under several feet of water.³⁰ But while even such convoluted claims may win this or that lawsuit or regulatory battle, ecological logic no longer seems unassailable. Once environmentalism can no longer put its concerns beyond politics, its posture becomes defensive rather than confident; its momentum is lost; and a heretofore impregnable regulatory regime becomes open to political challenge.

The trajectory of environmental legislation since 1990 illustrates the significance of this change in the terms of the debate. In 1990, Congress enacted the Clean Air Act Amendments, already mentioned as the most convoluted environmental statute on record. Although no national crisis or political necessity compelled their enactment, the amendments passed with the support of the Bush administration and of both parties in Congress. Only two years later, the environmental movement euphorically greeted the arrival of the 103rd Congress and of a firmly pro-environmentalist Democratic administration (the first after twelve years of what environmentalists viewed as a Republican drought). Far from living up to the environmental movement's high expectations, however, the 103rd Congress turned into the worst environmentalists had seen in over two decades. Republican and Democratic representatives began to attach amendments requiring property rights protections, prohibitions on unfunded mandates, and cost-benefit standards—the aforementioned "unholy trinity"—to pending reauthorization bills. The amendments picked up increased support with each consecutive vote. Eventually, environmental groups were forced to adopt a kill strategy to prevent the enactment of legislative amendments they viewed as a retreat from the existing environmental commitments.³¹ As a result, Congress failed to pass any significant environmental legislation.

Having assumed control over both houses of Congress in 1994, the Republican majority moved aggressively to advance the regulatory reform agenda. Intense political resistance forced the Republican lead-

---

30. Confronted with the NRDC's claim to this effect, Judge Douglas Ginsburg conceded that the allegations of a complete destruction of California "make out an injury indeed." Judge Ginsburg, however, was unwilling to extend standing to "anyone with the wit to shout 'global warming' in a crowded courthouse." City of Los Angeles v. NHTSA, 912 F.2d 478, 483–484 (D.C. Cir. 1990).

31. John H. Cushman, Jr., "Environmental Lobby Beats Tactical Retreat," *New York Times,* March 30, 1994. The environmentalist kill strategy was outlined in a memorandum, authored by an NRDC lobbyist, that was leaked to the press and published in the *BNA National Environment Daily,* March 16, 1994.

ership to retreat; to date, a diluted version of unfunded mandates legislation remains the only part of the unholy trinity to have been enacted. In this light, and for additional reasons discussed below, serious questions remain about the scope and timing of environmental law reform. But the terms of the debate have shifted too dramatically to leave much doubt about the general direction. The legislative agenda for property rights and risk-benefit standards has a great deal of intuitive plausibility and public support. (Enhanced protection for property rights was an element of the House Republicans' Contract with America, which was extensively pretested in polls and focus groups.)[32] Comprehensive command-and-control schemes no longer have political appeal; a replay of the 1990 Clean Air Act Amendments is out of the question. After decades in the political wilderness, libertarian advocacy groups are finding a receptive audience on Capitol Hill; environmental groups are fighting a defensive war.

It will be difficult to arrest this momentum. As noted, the harm-based doctrines that spell environmentalism's demise ultimately point toward private orderings and principled restraints on environmental regulation, and Congress has already indicated a willingness to follow this path further than the courts have been prepared to go. Proposed legislative standards of "reasonableness" are far more stringent than those suggested by *Corrosion Proof Fittings* or *International Union*. Property rights bills introduced by the Republican leadership in the House and the Senate closely track the common-law rationales of *Lucas*—and extend the protection *Lucas* offers for total wipeouts to partial takings in excess of one-third of the property value. Such proposals, now widely denounced as extreme, may make more headway in years to come. As environmentalism loses its resonance and residual force, legislators may find it as difficult to vote against property rights or for regulatory standards that kill people as they once found it to vote against clean air.

A decade or so hence, a historian of twentieth-century social movements may conclude that environmentalism's demise circa 1992 was a foregone conclusion. Eventually, the ecological regime *had* to

---

32. See Ed Gillespie and Bob Schellhas, eds., *Contract with America: The Bold Plan by Rep. Newt Gingrich, Rep. Dick Armey and the House Republicans to Change the Nation* (New York: Times Books, 1994). The Contract with America includes the Job Creation and Wage Enhancement Act, which will decrease government regulation and "protect individual Americans from overzealous federal regulators . . . [and] make sure that private property cannot be taken away without just compensation." Ibid., p. 125.

collapse under its contradictions—the mounting costs, the disappointing results, the experts' growing exasperation, the expanding regulatory reach from "corporate polluters" to municipalities, dry cleaners, and small land owners, which generated public resistance, changed the tone of the news coverage, and spawned a broad-based and effective property rights movement. But such broad, slow-moving currents pick up speed and punch only when they are directed into narrower channels, and the courts provided such a channel. The experts' consensus against command-and-control regulation failed to change the debate—and, in fact, left Congress and the EPA thoroughly unimpressed[33]—until it found its way into appellate decisions. The property rights movement, like any movement, needed a symbol and a rallying point, and *Lucas* was that symbol. From a broad historical perspective, the judiciary's role in channeling and articulating general discontents may seem marginal against the larger background of an inevitable collapse. But political regimes rarely collapse under their own weight; more often, they tumble when someone removes bricks from the ideological foundations. We should find the perpetrators before the fog of historical inevitability settles on the crime scene. The available evidence puts the courts high on the list of suspects.

Politics as Second Best

In 1984, R. Shep Melnick concluded a trenchant article on the use and abuse of symbolic, absolutist environmental laws with the observation that "our symbols are more in need of reform than our practices."[34] By confining transcendental values to interest group politics, the courts have mended our symbols. In so doing, they have enabled Congress to mend our practices. Alas, Congress may not live up to this challenge. The cheerful scenario just outlined—a progressive legislative move

---

33. Proposals for "market-based" or "incentive-based" reforms began to dominate the academic debate more than a decade ago. Among the most ardent proponents were some of the original architects of environmental regulation and, a decade later, environmental think tanks and advocacy groups such as Resources for the Future and the Environmental Defense Fund. See, for example, Bruce A. Ackerman and Richard B. Stewart, "Reforming Environmental Law," *Stanford Law Review*, vol. 37 (1985), p. 1333. For depressing accounts of how and why very little ever became of such reforms see Robert W. Hahn and Gordon L. Hester, "Where Did All the Markets Go?" *Yale Journal of Regulation*, vol. 6 (1989), p. 109; and Richard A. Liroff, *Reforming Air Pollution Regulation: The Toil and Trouble of EPA's Bubble* (Washington, D.C.: Conservation Foundation, 1986).

34. Melnick, "Pollution Deadlines and the Coalition for Failure," p. 134.

toward greater restraints on environmental regulation—conceals darker possibilities.

Political movements and organizations are stubborn beasts; they tend to stick around long after the faith on which they were founded has become discredited. Like a religion that has lost its core, environmentalism without the paradigm of complexity lacks coherence and vitality. But its high priests are not so easily discouraged; they continue to proselytize and to profess their faith even if it no longer wins converts. Similarly, government programs, laws, regulations, and agencies clutter the landscape long after the passions or needs that prompted their creation have subsided or disappeared.

Thus, although environmentalism is a spent force, its residual effects may well continue to hinder the reform agenda in the near future. More ominously, environmentalism threatens to leave in its wake laws and institutional arrangements that, deprived of their ideological underpinnings, turn into a vast game of rank redistribution. (In its sheer scope, this regime would be second only to the tax code and one or two entitlement programs.) The principal support for such a system of environmental redistribution would come, not from environmental advocacy groups, but from the regulated industries. Some industries—such as hazardous waste companies and ethanol producers—depend for their livelihood on stringent environmental standards; other industries will seize on such standards to drive competitors out of business; still others will want to preserve existing standards to recoup the sunk costs of investments made long ago in response to legislative mandates. Environmental advocates, for their part, will prefer ecological gestures to nothing at all. They will cut deals with corporate interests and, when called on, put their tarnished environmental icons on display—much as the ghosts of family farmers periodically haunt the halls of Congress to urge another round of subsidies for agricultural conglomerates.

Trends in this direction, evident for some time,[35] may well overwhelm the regulatory reform agenda. Property rights protections and cost-benefit requirements will be enacted in one form or another. But they may not be enough. If regulators can no longer confiscate private property in one or two steps, they will learn how to do it in five. If the benefits of oil drilling in a wildlife refuge would exceed the costs at

---

35. For a brief account, see Michael S. Greve, "Environmental Politics without Romance," in Michael S. Greve and Fred L. Smith, eds., *Environmental Politics: Public Costs, Private Rewards* (New York: Praeger, 1992). Several chapters in this volume provide detailed accounts of special interest policies in environmental disguise.

arbitrary value x per caribou, arbitrary value $x + 1$ may well produce the desired result. The unholy trinity may turn into the NEPA of environmental reform: a powerful symbol at first, a victim of entrenched political forces in the end.

One cannot look to the courts to be of much help in foreclosing such results. Certainly, the current doctrines could be expanded: *Lucas* can be extended by manipulating the "relevant parcel" or the "sticks" in the bundle of property under the takings analysis; the more particularized the definition of individual parcels, the more readily a complete (and therefore compensable) taking of one parcel or another will be found. De facto harms under the standing analysis can be defined more narrowly. Criteria of reasonableness can be applied more aggressively, and, in interpreting legislative language, courts have considerable maneuvering room before reaching the conclusion that Congress unmistakably intended to be unreasonable.[36] Even so expanded, however, the doctrines allow comfortable room for interest-group bargains.

Following the doctrinal logic *beyond* this point would entail a major conflagration with Congress because it would foreclose the redistributive games that are the legislature's ordinary mode of operation. The courts have not risked such a confrontation in six decades. The idea that they might do so over the environment is almost as absurd as the idea of the supremacy of ecological values—and for the same reason: the environment is simply not central to the nation's fortune and self-understanding.[37]

In any event, a further move toward private, common-law orderings could not be confined to environmental regulation; it would entail a wholesale attack on the administrative state. An extension of *Lucas* to the context of partial takings would compel compensation for just about any environmental regulation that exceeds the boundaries of nuisance law. This includes the Endangered Species Act, wetlands regulations under the Clean Water Act, and large portions of the Surface Mining Control and Reclamation Act, as well as myriads of state and local laws and regulations. A further extension beyond the land use area to commercial transactions would bring down the entire regu-

---

36. Edward W. Warren and Gary E. Marchant, "More Good than Harm; A First Principle for Agencies and Reviewing Courts," *Ecology Law Quarterly*, vol. 20 (1993), p. 431.

37. See Rosenberg, *The Hollow Hope*, pp. 286–87 (poll data show substantial public support for environmental protection, but the percentage of citizens listing the environment as among the most important problems facing the country never exceeded 10 percent).

latory state.³⁸ Similarly, under *Defenders of Wildlife,* Congress may still encourage interest groups—including environmental interest groups—to run into court when statutes enacted for their benefit have not been administered to their satisfaction. Standing rules that confined private litigation against the government more clearly to claims of common-law rights, however, would prevent just about any regulatory beneficiary from challenging regulatory practices.³⁹ And so, finally, with the reasonableness standard suggested in *Corrosion Proof Fittings* and *International Union:* while the requirement that regulations should do no more harm than good (in terms of health and safety) may seem unduly modest, a more demanding rule would threaten the entire regulatory edifice.⁴⁰

The courts could not possibly sustain such an assault without the complicity or at least the tacit consent of Congress. This, in turn, would

---

38. So extended, the logic of *Lucas* is essentially the logic of Richard Epstein's *Takings: Private Property and the Power of Eminent Domain* (Cambridge: Harvard University Press, 1985), which concludes that the takings clause poses an insuperable constitutional obstacle to the entire New Deal, just about all modern welfare and environmental regulation, and, for good measure, the progressive income tax.

39. Even objects of regulation—that is, members of regulated industries—would encounter standing problems under standing rules modeled more closely on the common law. A broadcaster, for example, might lack standing to challenge the Federal Communications Commission's unlawful award of a license to a competitor: after all, there is no common-law right to freedom from competition. The oddity of this result points again to the close connection between common-law property rights and standing rules. Under a common-law regime, competitor *C* is a mere bystander to an agreement or a legal dispute between *A* and *B* and therefore has no substantive right or cause of action. Strictly speaking, there is no need for standing rules. Once this robust notion of rights has been abandoned and the legitimacy of administrative regimes has been conceded, however, competitors are no longer bystanders but members of the regulated industries. They must have standing, and standing rules must ultimately be based on common-law *analogs.* See Sunstein, "Standing and the Privatization of Public Law," pp. 1433–38. The Supreme Court permitted competitor standing long before the New Deal: Chicago Junction Case, 264 U.S. 258, 262–69 (1924).

40. This is true even of tests that fall well short of the rule, suggested in note 23 above, that environmental standards in derogation of common-law arrangements should be narrowly construed. Most economists, for example, would defend a requirement that rules and standards must produce net economic benefits (of all kinds, not simply in terms of health or safety) *at the margin,* not merely on average. Existing legislative health and technology standards often fail this test.

require a public consensus on a drastic curtailment of interest-group politics—in effect, a public consensus to repeal the New Deal. One may forgive the courts if they fail to see such an agreement on the horizon.

To be sure, there is a widespread disgust with interest-group politics and government meddling. This sentiment may eventually gel into a libertarian consensus. For the time being, however, public cynicism and antigovernment sentiments exist alongside, and in some ways arise from, the great expectations the public continues to have of government and its capacity to address urgent social problems. This ambivalence translates, not into a consensus, but into a rancorous political debate and a series of uneasy and unsatisfying compromises.

In the environmental area, a huge gap between antiregulatory sentiments and wildly inflated expectations is reflected in the contentious congressional debate over the unholy trinity. The elements of that trinity are extremely popular: large majorities of voters think that private property should be protected, that regulations should do more good than harm, and that Congress should fund what it mandates. But equally large majorities of voters agree that environmental standards cannot be high enough.[41] The voters favor balance; at the same time, they oppose an "environmental rollback" that would dismantle huge chunks of the existing environmental apparatus. They have yet to discover that they cannot have it both ways.

Inconsistent preferences are not perplexing; they are the stuff of life. We all want fiscal discipline and lower taxes—and higher spending in our own bailiwicks. We all want to control our neighbors—without being subject to similar controls. The point of the common law is to teach that we cannot have it both ways: there is no free lunch, no reward without risk, no right without a reciprocal obligation. From this harshness of markets, we persistently seek escape into politics, which holds out the promise of a lunch that, if not exactly free, is paid by our neighbors—if only we invest sufficient time, effort, and money to defeat our neighbors' designs to have *us* foot the bill.

But the chasm between our professed *environmental* convictions and our sense of balance is still deeper. When we argue about social security or welfare reform or even affirmative action, we understand that any solution will entail costs and benefits. We argue about the magnitude of those costs and benefits and about their distribution. But we do not contest the principle that we should do more good than harm in some tangible sense. While we disagree and argue about the

---

41. For recent polling data confirming the public's ambivalence, see Adler, *Environmentalism at the Crossroads*, pp. 144–46.

value of things and while we want more for ourselves than for others, we do not pretend that *no* amount of good x will compensate for a little less of good y.

Environmentalism, in contrast, attempts to put *its* concerns beyond such pedestrian calculations. It elevates them into the realm of values and relegates all else to the sordid world of political bargaining. The environment is priceless. It is our collective possession, and any demand for some sort of political accommodation, some sort of comparison of costs and benefits—not to mention the chaos of private arrangements—is effrontery. Between absolute values and real-world concerns, no middle ground exists. This mindset has left traces on American politics that are not easily erased.

For this reason the judicial demise of the ecological paradigm merits a cheer, perhaps two. The courts have reminded us of the costs of an unconstrained political process and reined in its worst excesses. Most important, the courts have reduced our expectations and reminded us that all values are *somebody's* values. Not much is said these days for interest-group politics. Perhaps not much *can* be said for it. A system that promises to each far more than it can deliver to any tends to produce strife and cynicism, and the attendant costs have become quite obvious. But so long as we resist the discipline of private orderings, it is best that interests appear on stage *as interests* instead of masquerading as values. Our politics will be less edifying but more realistic. In time, we may yet come to realize that things that once seemed hopelessly complex are in the end quite manageable, that everything has its price, and that the risks of ordinary life may pale against the risks of politics.

Index

Administrative Procedures Act (1946), 45, 47–48
Agency rulings
    to avoid trade-offs, 78–84
    condition for hard-look review of, 65
    reasonableness criterion, 66
    set-aside and remanded by judicial decision, 69–72
Allen v. Wright (1984), 54–55
American Textile Manufacturers v. Donovan (1981), 70–71, 73–74

Babbitt v. Sweet Home Chapter (1995), 67, 83–84
Beachfront Management Act, South Carolina, 23–24, 35–36
Benzene case. See *Industrial Union Dept., AFL-CIO v. American Petroleum Institute* (1980)
Blackmun, Harry
    dissent in *Defenders of Wildlife*, 61, 62
    dissent in *Lucas*, 35, 117
Breyer, Stephen, 77 n. 42; 88 n. 7; 91 n. 12; 92 nn. 13–14, 16; 97 n. 30

Calvert Cliffs Coordinating Committee v. Atomic Energy Commission (D.C. Cir. 1971), 1
Capture theory
    critique of, 102–3
    different interpretations of standing to sue, 44–45
    environmental values concept tied to, 65
    packaged with ecological paradigm, 67
CEI II. See *Competitive Enterprise Institute v. National Highway Traffic Safety Administration* (1992)

Chevron v. NRDC (1985)
    circumstances for policy change in, 104
    deference under, 94, 98, 104, 123
    effect of litigation on policy, 105
    reasonable interpretation, 66–67
Citizen suits
    connection to regulatory failures, 99
    effect of, 101–2
    effect on regulatory obsolescence, 104
    invention and premise of environmental, 126–27
    lack of judicial scrutiny of standing, 99
Civil rights
    granting of standing in cases of, 57–59
    law, 16–17
Clean Air Act (1970)
    1990 amendments, 3, 114, 130–31
    beginning of environmental era, 2
    provision for citizen suit, 42, 127
Commoner, Barry, 4
Common law
    as basis for injury-in-fact test, 122
    effect of New Deal legislation on doctrines of, 8–9
    harm-based boundaries of, 85
    harm-based doctrines, 89–90
    author's meaning of, 115
    ideological argument against, 115–16
    in *Lucas*, 24–25, 40, 123
    of nuisance, 31–40
    objectivity and value-free nature of, 40
    opposition to interest-group pluralism, 110

property rights under, 26, 39, 117
rules to manage complexity,
116–17
view of public-law scholars,
12–13
Common-law rights
environmentalism view of, 1–2
premise in *Defenders of Wildlife*,
60
See also Private property; Property rights
Competition, environmental movement, 101
Competitive Enterprise Institute v. National Highway Traffic Safety Administration (1992)
courts' interpretation of congressional intent, 72, 75
courts' perception of regulatory failure, 76–83
decision in, 80
deference in judicial review, 96
remanded rulings from, 69–72, 98
substantive review practiced in, 82–84, 94, 98–99
Congressional intent, 72–76
Conservation Law Foundation of New England v. Reilly (1991), 50–51
Corrosion Proof Fittings v. EPA (1992)
courts' focus on trade-offs and costs, 96
courts' interpretation of congressional intent, 72–75
courts' perception of regulatory failure, 76–83
deference in judicial review, 95
reasonableness standard in, 135
set-aside and remanded rulings from, 69–72
substantive review of regulations, 94, 98–99
Cotton Dust case. *See American Textile Manufacturers v. Donovan* (1981)
Courts
application of reasonableness standard, 97
environmentalist view of access to, 5–6
interpretation of congressional intent, 72–76

perception of regulatory failure, 76–83
rejection of ecological paradigm, 3, 18–20, 42–43, 63, 84
role in demise of environmentalism, 109–10, 131–32
role in environmental policy process, 124–26
role in policy process, 124–26
shift to harm-based doctrine, 91
See also Judicial decisions; Judicial review; Supreme Court

Defenders of Wildlife. See Lujan v. Defenders of Wildlife (1992)
Discretion, executive, 65
Dolan v. Tigard (1994), 30
Douglas, William O., 45

Ecological paradigm
basic premises and source of, 1, 5–7, 110–11
challenge to ideas of property, 24
connection between takings and standing under, 61–62
differences from New Deal legislation, 10–11
influence on American law, 8
interpretations of Scalia and Kennedy, 62–63
purpose of, 107–8
rejection by courts, 23, 42–43, 63, 84, 128
rise and demise of, 22, 46, 85–86, 124
role of values in statutory interpretation, 64–66
as source of regulatory failures, 20–21, 89
See also Environmentalism
Ehrlich, Paul, 111–12
Endangered Species Act (1973), 4, 90
Enforcement
effect of environmental, 91–92
goals of environmental plaintiffs, 100
Environmentalism
cause of demise, 131–32
demise of, 108
as ideology, 1–2, 113–15
inferences and effect of, 111–15

premise of interconnectedness, 3–5, 43, 85
  See also Ecological paradigm
Environmental movement
  citizen suits as tools of, 126–27
  declining membership in, 109
  institutional makeup, 100–101
Environmental policy
  coordination problems in, 91–92
  effect of litigation on, 105
  interest-group politics in making of, 67
  National Environmental Policy Act, 2, 46, 126–27
  role of courts in, 124–26
  See also Congressional intent; Legislation; Legislation, environmental; Regulation
Executive Power: Standing, Separation of Powers, and Article II, 54–61

Federal Land Policy and Management Act (1976), 46

Ginsburg, Douglas, 97, 98

Hard-look review
  arguments for, 93
  criticism of, 93–94
  departure from, 73
  ecological, 72
  purpose of, 65, 95, 104, 127
  reasoned decision-making requirement, 77–78
  tendency to produce regulatory obsolescence, 104
Harm-based doctrines
  common-law–like conceptions, 122–24
  to correct and replace ecological paradigm, 85–89, 110, 118, 121
  to correct regulatory failures, 89–90
  courts' shift to, 91
  entitlements and interest-group politics in, 121
  See also Injury-in-fact test
Harms
  Disconnection of standing to sue barriers from, 42
  environmentalist definition, 6–7

as standard in injury-in-fact cases, 45
Ideology
  attacks on common law, 115–16
  embodied in case law, 107
  environmentalism as, 1–2, 113–15
  See also Ecological paradigm; Environmentalism
Industrial Union Dept., AFL-CIO v. American Petroleum Institute (1980), 70–71, 73–74
Injury
  definition in *Defenders of Wildlife*, 55–61
  environmentalism's definition, 6
Injury-in-fact test
  as basic standing requirement, 45
  purpose of and basis for, 122
  used by environmental litigants, 46
Interconnectedness principle, 3–5, 15, 43, 85
Interest-group politics
  conflict with common-law regime, 118–24
  effect on regulatory process, 102
  in harm-based doctrines, 121
  influence on New Deal legislation, 9
  policy-making role, 67
  as replacement for private orderings, 118
International Union, UAW v. Occupational Safety and Health Administration (1991)
  courts' interpretation of congressional intent, 72–75
  courts' perception of regulatory failure, 76–84
  deference in judicial review, 95–96
  description of case, 70
  reasonableness standard in, 135
  set-aside and remanded rulings from, 69–72, 97–98
  substantive review of regulations, 94–95, 98–99

Judicial decisions
  during environmental era (1970s and 1980s), 8

INDEX

environmentalist logic (1970s and 1980s), 2
reflecting substantive review, 68–72
related to property owners, 18–20
Judicial review
under Administrative Procedures Act (1946), 45, 47–48
effects on regulatory obsolescence, 104
of environmental, health, and safety regulation, 66
suggested limits for, 93
using private-ordering base line, 123
See also Hard-look review; Substantive review

Kennedy, Anthony
concurrence in *Lucas*, 36–38, 41
opinion in *Defenders of Wildlife*, 62

Land withdrawal review program, 46–52
Law of unintended consequences, 21
Legal interest test, 45
Legislation
environmental backlash, 109
proposed standards of reasonableness, 131
public-law notion of, 13
Legislation, environmental
provisions for citizen suits, 42, 51, 127
regulation of private property, 8
Litigation
effect on environmental policy, 105
under NEPA, 126
purpose and impact of industry, 100
purpose of beneficiary, 100
standing to sue in, 46
Lochner v. New York (1905), 118
Lucas v. South Carolina Coastal Council (1991)
Blackmun's dissent in, 35, 117
case described, 23–24
common-law rationale as model, 131

dissenters on common law of nuisance, 39
interpretation of property rights in, 123, 128–29
interpretation of takings law in, 23–25, 30–41, 91
Kennedy's concurrent opinion, 36–38
Scalia's opinion in, 23–24, 30–32, 38, 41, 62, 128
Lujan v. Defenders of Wildlife (1992)
denial of standing to environmental litigants, 46, 52–63
interest groups under, 123
limits to citizen standing, 99
standing analysis of, 121–22
Lujan v. National Wildlife Federation (1990)
environmentalists' loss of preferred status, 46–52
land withdrawal review program in, 46–52
Scalia's opinion in, 47, 49–50

Melnick, R. Shep, 103, 127, 132

National Environmental Policy Act, 1969 (NEPA)
beginning of environmental era with, 2
case related to violation of, 46–52
requirement for environmental impact statements, 126–27
National Wildlife Federation. *See Lujan v. National Wildlife Federation* (1990)
New Deal legislation
cartel arrangements under, 10, 119
differences from ecological paradigm, 10
effect on common-law doctrines, 8–9, 116, 120
standing granted under, 45
Nollan v. California Coastal Commission (1987), 29, 32

Obsolescence, regulatory
causes of, 103–4
defined, 89
Delaney clause as, 105–6

142

effect of citizen suits on, 105
effect of judicial review on, 104–5
O'Connor, Sandra Day, 23

Private property
  common-law tradition, 4
  conditions for government taking of, 25–26
  under environmentalist logic, 5
  requires protection, 117
  under takings clause, 23–24
Property owners
  expansion of Fifth Amendment protection, 18–19
  focus in *Lucas* decision, 38–40
  position under environmental regulations, 28
  protection under common law, 39
  See also Private property
Property rights
  based on common-law principles in *Lucas*, 30 40
  under common law, 26, 60–61, 117
  environmentalism's analysis of, 6, 85
  exclusive control as essence of, 60–61
  interpretation in *Lucas*, 123, 129
  link between law, regulation, and outcomes, 90–91
Property rights movement, 109
Public-choice theory, 68–72, 98
Public-law theory
  criticism of injury-in-fact criterion, 45
  with emergence of environmentalism, 12
  environmental values in, 13–14
  influence of, 15–16
  position on civil rights law, 16–17
  view of common law, 13

Rabkin, Jeremy A., 7
Reasonableness standard
  in *Corrosion Proof Fittings*, 135
  court application of, 97
  imputation of, 72
  in *International Union*, 74, 135
  in interpretation of regulations, 73–74
  presumption of, 72–73, 95–96
  proposed, 131
Regulation
  pursuit of takings through, 25–26
  reasonableness standard in interpretation of, 73–74
  regulatory failure, 76–84
  value-driven, 21
  See also Hard-look review; Judicial review; Substantive review
Regulation, environmental
  coordination with substantive review, 95
  criticism of current regime, 88–89
  limitations of, 107
  logic of environmental ideology in, 108
  permanent pursuit of values in, 11
  position of property owners under, 28
  problems in coordination, 91–92
  proposed curtailment, 3
  from public-choice theory perspective, 68–72, 98
  public values in, 13
Rehnquist, William, 23
Rights
  establishment of environmental rights, 3
  of individual under common law, 115
  protection under common law, 58
  See also Property rights
Risk, marginal
  agency avoidance of trade-offs with, 79

Sax, Joseph L., 24, 41
Scalia, Antonin
  arguments for respect for agency authority, 95
  argument related to substantive review, 84 n. 64
  dissent in *Sweet Home Chapter*, 67, 84

INDEX

"Doctrine of Standing as an Essential Element of the Separation of Powers, The," 1, 53, 10
  opinion in *Defenders of Wildlife*, 53–55, 56, 62, 106
  opinion in *Lucas*, 23–24, 30–32, 38, 41, 62, 129
  opinion in *National Wildlife Federation*, 47–49, 52, 106
  opinion in *Nollan*, 29, 32
  position on judiciary's policy-making role, 66
  position on reasonable interpretation of statute, 66
Separation-of-powers principle
  in *Defenders of Wildlife* decision, 53–58
  in *National Wildlife Federation* decision, 52
Standing rules, harm-based, 44–45
Standing to sue
  analysis in *Defenders of Wildlife*, 52–62, 121–22
  analysis in *Lujan v. National Wildlife Federation*, 46–52
  court denial in *Allen v. Wright*, 54
  definition, 43
  denial of standing in environmental cases, 46–55
  effect of citizen, 127
  environmentalism's view of, 5–6
  in environmental litigation, 46
  under injury-in-fact test, 45
  interpretation in environmentalist's view, 43–46
  under New Deal legislation, 10
  real harms as requisite for, 115
  See also Citizen suits
Stewart, Richard B., 21
Substantive review
  agency leeway with, 104
  of agency regulations, 77
  argument for, 82–84, 95
  characterization of, 68
  conditions for deference to agency determination, 83–84
  cost-benefit analysis, 74–75, 79–80, 96
  decisions reflecting style of, 68–72, 95–96, 98
  description of style, 68

presumption of reasonableness of legislation, 95
  in *Sweet Home Chapter*, 83–84
  tools for, 83
Sunstein, Cass, 18, 56, 57, 60, 88–89, 96, 102, 121–22
Supreme Court
  cases in which standing denied, 46–47
  decision in *Babbitt v. Sweet Home Chapter*, 67
  decision in *Lucas*, 23–25, 30–40, 91
  decision related to takings in *Nollan*, 29
  denial of standing, 54, 56–57
  granting of standing, 46, 56–57
  position on environmental issues, 29
Sweet Home Chapter. *See Babbitt v. Sweet Home* chapter (1995)

Takings
  common-law conception in analysis of, 24–25
  pursued through regulation, 25–26
  requiring compensation, 25
  Supreme Court decisions related to, 29–41
Takings clause, Fifth Amendment
  historic purpose of, 28
  interpretation of environmentalism, 26–28
  provision related to private property, 23
  public needs and private rights balanced in, 25
Takings law
  common-law focus in *Lucas*, 40
  environmental issues in, 29
  interpretation in *Nollan*, 29
Thomas, Clarence, 23

"Unholy Trinity," 109, 130, 134, 136

Values
  barriers to imposition of, 121
  demise of environmentalism, 2, 20–21

144

enforcement by environmental plaintiffs, 100
in environmental impact statements, 126
in environmental statutes, 64
lack of exclusivity in public, 61
limits on environmental, 115, 118
transcendence of environmental, 112–15

Value-oriented jurisprudence, 120
Values concept, 65

White, Byron, 23
Williams, Stephen, 68–70, 76, 78, 82, 97, 98
Willingness-to-pay criterion, 91
Wright, Skelly, 104

About the Author

MICHAEL S. GREVE is a cofounder and, since 1989, the executive director of the Center for Individual Rights, a public-interest law firm in Washington, D.C. He has taught political science courses at Hunter College, John Jay College, and Cornell University.

Mr. Greve's numerous articles on environmental law and policy have appeared in law reviews, scholarly journals, and newspapers and journals such as *The American Enterprise* and *The Public Interest*. He is a coeditor, with Fred L. Smith, of *Environmental Politics: Public Costs, Private Rewards* (Praeger, 1992). Mr. Greve is a director of the Competitive Enterprise Institute and an adjunct scholar of the American Enterprise Institute.

He received his M.A. and Ph.D. from Cornell University.

Board of Trustees

Wilson H. Taylor, *Chairman*
Chairman and CEO
CIGNA Corporation

Tully M. Friedman, *Treasurer*
Hellman & Friedman

Edwin L. Artzt
Chairman of the Executive Committee
The Procter & Gamble
 Company

Joseph A. Cannon
Chairman and CEO
Geneva Steel Company

Dick Cheney
Chairman, President, and CEO
Halliburton Company

Albert J. Costello
President and CEO
W. R. Grace & Co.

Harlan Crow
Managing Partner
Crow Family Holdings

Christopher C. DeMuth
President
American Enterprise Institute

Malcolm S. Forbes, Jr.
President and CEO
Forbes Inc.

Christopher B. Galvin
President and COO
Motorola, Inc.

Harvey Golub
Chairman and CEO
American Express Company

Robert F. Greenhill
Chairman
Greenhill & Co. LLC

M. Douglas Ivester
President and COO
The Coca-Cola Company

Martin M. Koffel
Chairman and CEO
URS Corporation

Bruce Kovner
Chairman
Caxton Corporation

Kenneth L. Lay
Chairman and CEO
ENRON Corp.

Marilyn Ware Lewis
Chairman
American Water Works Company, Inc.

Alex J. Mandl
President and COO
AT&T

Craig O. McCaw
Chairman and CEO
Eagle River, Inc.

Paul H. O'Neill
Chairman and CEO
Aluminum Company of America

George R. Roberts
Kohlberg Kravis Roberts & Co.

John W. Rowe
President and CEO
New England Electric System

Edward B. Rust, Jr.
President and CEO
State Farm Insurance Companies

James P. Schadt
Chairman and CEO
Reader's Digest Association, Inc.

John W. Snow
Chairman, President, and CEO
CSX Corporation

William S. Stavropoulos
President and CEO
The Dow Chemical Company

The American Enterprise Institute for Public Policy Research

Founded in 1943, AEI is a nonpartisan, nonprofit, research and educational organization based in Washington, D.C. The Institute sponsors research, conducts seminars and conferences, and publishes books and periodicals.

AEI's research is carried out under three major programs: Economic Policy Studies; Foreign Policy and Defense Studies; and Social and Political Studies. The resident scholars and fellows listed in these pages are part of a network that also includes ninety adjunct scholars at leading universities throughout the United States and in several foreign countries.

The views expressed in AEI publications are those of the authors and do not necessarily reflect the views of the staff, advisory panels, officers, or trustees.

James Q. Wilson
James A. Collins Professor
 of Management
University of California
 at Los Angeles

Officers

Christopher C. DeMuth
President

David B. Gerson
Executive Vice President

Council of Academic Advisers

James Q. Wilson, *Chairman*
James A. Collins Professor
 of Management
University of California
 at Los Angeles

Gertrude Himmelfarb
Distinguished Professor of History
 Emeritus
City University of New York

Samuel P. Huntington
Eaton Professor of the
 Science of Government
Harvard University

D. Gale Johnson
Eliakim Hastings Moore
 Distinguished Service Professor
 of Economics Emeritus
University of Chicago

William M. Landes
Clifton R. Musser Professor of
 Economics
University of Chicago Law School

Sam Peltzman
Sears Roebuck Professor of Economics
 and Financial Services
University of Chicago
 Graduate School of Business

Nelson W. Polsby
Professor of Political Science
University of California at Berkeley

George L. Priest
John M. Olin Professor of Law and
 Economics
Yale Law School

Murray L. Weidenbaum
Mallinckrodt Distinguished
 University Professor
Washington University

Research Staff

Leon Aron
Resident Scholar

Claude E. Barfield
Resident Scholar; Director, Science
 and Technology Policy Studies

Cynthia A. Beltz
Research Fellow

Walter Berns
Resident Scholar

Douglas J. Besharov
Resident Scholar

Robert H. Bork
John M. Olin Scholar in Legal Studies

Karlyn Bowman
Resident Fellow

John E. Calfee
Resident Scholar

Lynne V. Cheney
W. H. Brady, Jr., Distinguished Fellow

Dinesh D'Souza
John M. Olin Research Fellow

Nicholas N. Eberstadt
Visiting Scholar

Mark Falcoff
Resident Scholar

John D. Fonte
Visiting Scholar

Gerald R. Ford
Distinguished Fellow

Murray F. Foss
Visiting Scholar

Diana Furchtgott-Roth
Assistant to the President and Resident
 Fellow

Suzanne Garment
Resident Scholar

Jeffrey Gedmin
Research Fellow

Patrick Glynn
Resident Scholar

Robert A. Goldwin
Resident Scholar

Robert W. Hahn
Resident Scholar

Thomas Hazlett
Visiting Scholar

Robert B. Helms
Resident Scholar; Director, Health
 Policy Studies

Glenn Hubbard
Visiting Scholar

Douglas Irwin
Henry Wendt Scholar in Political
 Economy

James D. Johnston
Resident Fellow

Jeane J. Kirkpatrick
Senior Fellow; Director, Foreign and
 Defense Policy Studies

Marvin H. Kosters
Resident Scholar; Director,
 Economic Policy Studies

Irving Kristol
John M. Olin Distinguished Fellow

Dana Lane
Director of Publications

Michael A. Ledeen
Resident Scholar

James Lilley
Resident Fellow; Director, Asian
 Studies Program

John H. Makin
Resident Scholar; Director, Fiscal
 Policy Studies

Allan H. Meltzer
Visiting Scholar

Joshua Muravchik
Resident Scholar

Charles Murray
Bradley Fellow

Michael Novak
George F. Jewett Scholar in Religion,
 Philosophy, and Public Policy;
 Director, Social and
 Political Studies

Norman J. Ornstein
Resident Scholar

Richard N. Perle
Resident Fellow

William Schneider
Resident Scholar

William Shew
Visiting Scholar

J. Gregory Sidak
F. K. Weyerhaeuser Fellow

Herbert Stein
Senior Fellow

Irwin M. Stelzer
Resident Scholar; Director, Regulatory
 Policy Studies

W. Allen Wallis
Resident Scholar

Ben J. Wattenberg
Senior Fellow

Carolyn L. Weaver
Resident Scholar; Director, Social
 Security and Pension Studies

A NOTE ON THE BOOK

*This book was edited by Ann Petty of the
publications staff of the American Enterprise Institute.
The index was prepared by Shirley Kessel.
The text was set in Palatino, a typeface designed by
the twentieth-century Swiss designer Hermann Zapf.
Coghill Composition Company, of Richmond, Virginia,
set the type, and Princeton Academic Press,
of Lawrenceville, New Jersey, printed and bound the book,
using permanent acid-free paper.*

The AEI PRESS is the publisher for the American Enterprise Institute for Public Policy Research, 1150 17th Street, N.W., Washington, D.C. 20036; *Christopher DeMuth,* publisher; *Dana Lane,* director; *Ann Petty,* editor; *Leigh Tripoli,* editor; *Cheryl Weissman,* editor; *Lisa Roman,* assistant editor (rights and permissions).